<u>Legal Notice</u>

BOOKS BY DR. STEVE WARNER FOR COLLEGE BOUND STUDENTS

28 New SAT Math Lessons to Improve Your Score in One Month
     Beginner Course
     Intermediate Course
     Advanced Course
New SAT Math Problems arranged by Topic and Difficulty Level
New SAT Verbal Prep Book for Reading and Writing Mastery
320 SAT Math Subject Test Problems
     Level 1 Test
     Level 2 Test
The 32 Most Effective SAT Math Strategies
SAT Prep Official Study Guide Math Companion
Vocabulary Builder
320 ACT Math Problems arranged by Topic and Difficulty Level
320 GRE Math Problems arranged by Topic and Difficulty Level
320 SAT Math Problems arranged by Topic and Difficulty Level
320 AP Calculus AB Problems
320 AP Calculus BC Problems
SHSAT Verbal Prep Book to Improve Your Score in Two Months
555 Math IQ Questions for Middle School Students
555 Advanced Math Problems for Middle School Students
555 Geometry Problems for High School Students
Algebra Handbook for Gifted Middle School Students

CONNECT WITH DR. STEVE WARNER

www.facebook.com/SATPrepGet800
www.youtube.com/TheSATMathPrep
www.twitter.com/SATPrepGet800
www.linkedin.com/in/DrSteveWarner
www.pinterest.com/SATPrepGet800
plus.google.com/+SteveWarnerPhD

320 ACT Math Problems arranged by Topic and Difficulty Level

320 Level 1, 2, 3, 4, and 5 Math Problems for the ACT

Dr. Steve Warner

Table of Contents

ACTIONS TO COMPLETE BEFORE YOU READ THIS BOOK

1. Purchase a TI-84 or equivalent calculator

It is recommended that you use a TI-84 or comparable calculator for the ACT. Answer explanations in this book will always assume you are using such a calculator.

2. Take a practice ACT from the Real ACT Prep Guide to get your preliminary ACT math score

Use this score to help you determine the problems you should be focusing on (see page 9 for details).

3. Claim your FREE bonuses

Visit the following webpage and enter your email address to receive an electronic copy of *The 32 Most Effective SAT Math Strategies* for FREE. You will also receive solutions to all the supplemental problems in this book, and an index of topics with a map to the "Real ACT Prep Guide."

www.thesatmathprep.com/ACTprmX.html

4. Like' my Facebook page

This page is updated regularly with ACT and SAT prep advice, tips, tricks, strategies, and practice problems.

www.facebook.com/SATPrepGet800

INTRODUCTION
THE PROPER WAY TO PREPARE

There are many ways that a student can prepare for the ACT. But not all preparation is created equal. I always teach my students the methods that will give them the maximum result with the minimum amount of effort.

The book you are now reading is self-contained. Each problem was carefully created to ensure that you are making the most effective use of your time while preparing for the ACT. By grouping the problems given here by level and topic I have ensured that you can focus on the types of problems that will be most effective to improving your score.

1. Using this book effectively

- Begin studying at least three months before the ACT
- Practice ACT math problems twenty minutes each day
- Choose a consistent study time and location

You will retain much more of what you study if you study in short bursts rather than if you try to tackle everything at once. So try to choose about a twenty-minute block of time that you will dedicate to ACT math each day. Make it a habit. The results are well worth this small time commitment.

- Every time you get a question wrong, **mark it off, no matter what your mistake**.
- Begin each study session by first redoing problems from previous study sessions that you have marked off.
- If you get a problem wrong again, **keep it marked off**.

Note that this book often emphasizes solving each problem in more than one way. Please listen to this advice. The same question is not generally repeated on any ACT so the important thing is learning as many techniques as possible.

7

Being able to solve any specific problem is of minimal importance. The more ways you have to solve a single problem the more prepared you will be to tackle a problem you have never seen before, and the quicker you will be able to solve that problem. Also, if you have multiple methods for solving a single problem, then on the actual ACT when you "check over" your work you will be able to redo each problem in a different way. This will eliminate all "careless" errors on the actual exam. Note that in this book the quickest solution to any problem will always be marked with an asterisk (*).

2. The magical mixture for success

A combination of three components will maximize your ACT math score with the least amount of effort.

- Learning test taking strategies that work specifically for standardized tests.
- Practicing ACT problems for a small amount of time each day for about three months before the ACT.
- Taking about four practice tests before test day to make sure you are applying the strategies effectively under timed conditions.

I will discuss each of these three components in a bit more detail.

Strategy: The more ACT specific strategies that you know the better off you will be. Throughout this book you will see many strategies being used. Some examples of basic strategies are "plugging in answer choices," "taking guesses," and "picking numbers." Some more advanced strategies include "identifying arithmetic sequences with linear equations," and "moving the sides of a figure around." Pay careful attention to as many strategies as possible and try to internalize them. Even if you do not need to use a strategy for that specific problem, you will certainly find it useful for other problems in the future.

Practice: The problems given in this book are more than enough to vastly improve your current ACT math score. All you need to do is work on these problems for about ten to twenty minutes each day over a period of three to four months and the final result will far exceed your expectations.

Let me further break this component into two subcomponents – **topic** and **level**.

Topic: You want to practice each of the five general math topics given on the ACT and improve in each independently. The five topics are **Number Theory**, **Algebra and Functions**, **Geometry**, **Probability and Statistics**, and **Trigonometry**. The problem sets in this book are broken into these five topics.

Level: You will make the best use of your time by primarily practicing problems that are at and slightly above your current ability level. For example, if you are struggling with Level 2 Geometry problems, then it makes no sense at all to practice Level 5 Geometry problems. Keep working on Level 2 until you are comfortable, and then slowly move up to Level 3. Maybe you should never attempt those Level 5 problems. You can get an exceptional score without them (over a 30).

Tests: You want to take about four practice tests before test day to make sure that you are implementing strategies correctly and using your time wisely under pressure. For this task you should use actual ACT exams such as those found in the third edition of "The Real ACT Prep Guide." Take one test every few weeks to make sure that you are implementing all the strategies you have learned correctly under timed conditions. Note that only the third edition has five actual ACTs.

3. Practice problems of the appropriate level

In this book ACT math questions have been split into 5 Levels. Roughly speaking, the ACT math section increases in difficulty as you progress from question 1 to questions 60. So you can think of the first 12 problems as Level 1, the next 12 as Level 2 and so on.

Keep track of your current ability level so that you know the types of problems you should focus on. If you are currently scoring around a 15 on your practice tests, then you should be focusing primarily on Level 1, 2, and 3 problems. You can easily raise your score by 4 points without having to practice a single hard problem.

If you are currently scoring about a 20, then your primary focus should be Level 2 and 3, but you should also do some Level 1 and 4 problems.

If you are scoring around a 25, you should be focusing on Level 2, 3, and 4 problems, but you should do some Level 1 and 5 problems as well.

Those of you at the 30 level really need to focus on those Level 4 and 5 problems.

If you really want to refine your studying, then you should keep track of your ability level in each of the five major categories of problems:

- **Number Theory**
- **Algebra and Functions**
- **Probability, Statistics and Data Analysis**
- **Geometry**
- **Trigonometry**

9

For example, many students have trouble with very easy geometry problems, even though they can do more difficult number theory problems. This type of student may want to focus on Level 1, 2, and 3 geometry questions, but Level 3 and 4 number theory questions.

4. Practice in small amounts over a long period of time

Ideally you want to practice doing ACT math problems ten to twenty minutes each day beginning at least 3 months before the exam. You will retain much more of what you study if you study in short bursts than if you try to tackle everything at once.

The only exception is on a day you do a practice test. You should do at least four practice tests before you take the ACT. Ideally you should do your practice tests on a Saturday or Sunday morning. At first you can do just the math section. The last one or two times you take a practice test you should do the whole test in one sitting. As tedious as this is, it will prepare you for the amount of endurance that it will take to get through this exam.

So try to choose about a twenty-minute block of time that you will dedicate to ACT math every night. Make it a habit. The results are well worth this small time commitment.

5. Redo the problems you get wrong over and over and over until you get them right

If you get a problem wrong, and never attempt the problem again, then it is extremely unlikely that you will get a similar problem correct if it appears on the ACT.

Most students will read an explanation of the solution, or have someone explain it to them, and then never look at the problem again. This is *not* how you optimize your ACT score. To be sure that you will get a similar problem correct on the ACT, you must get the problem correct before the ACT—and without actually remembering the problem.

This means that after getting a problem incorrect, you should go over and understand why you got it wrong, wait at least a few days, then attempt the same problem again. If you get it right, you can cross it off your list of problems to review. If you get it wrong, keep revisiting it every few days until you get it right. Your score *does not* improve by getting problems correct. **Your score improves when you learn from your mistakes.**

10

6. Check your answers properly

When you go back to check your earlier answers for careless errors *do not* simply look over your work to try to catch a mistake. This is usually a waste of time. Always redo the problem without looking at any of your previous work. Ideally, you want to use a different method than you used the first time.

For example, if you solved the problem by picking numbers the first time, try to solve it algebraically the second time, or at the very least pick different numbers. If you do not know, or are not comfortable with a different method, then use the same method, but do the problem from the beginning and do not look at your original solution. If your two answers do not match up, then you know that this a problem you need to spend a little more time on to figure out where your error is.

This may seem time consuming, but that's ok. It is better to spend more time checking over a few problems than to rush through a lot of problems and repeat the same mistakes.

7. Take a guess whenever you cannot solve a problem

There is no guessing penalty on the ACT. Whenever you do not know how to solve a problem take a guess. Ideally you should eliminate as many answer choices as possible before taking your guess, but if you have no idea whatsoever do not waste time overthinking. Simply put down an answer and move on. You should certainly mark it off and come back to it later if you have time.

8. Pace yourself

Do not waste your time on a question that is too hard or will take too long. After you've been working on a question for about 30 to 45 seconds you need to make a decision. If you understand the question and think that you can get the answer in another 30 seconds or so, continue to work on the problem. If you still do not know how to do the problem or you are using a technique that is going to take a long time, mark it off and come back to it later if you have time.

If you do not know the correct answer, eliminate as many answer choices as you can and take a guess. But you still want to leave open the possibility of coming back to it later. Remember that every problem is worth the same amount. Do not sacrifice problems that you may be able to do by getting hung up on a problem that is too hard for you.

11

9. Attempt the right number of questions

Many students make the mistake of thinking that they have to attempt every single ACT math question when they are taking the test. There is no such rule. In fact, most students will increase their ACT score by *reducing* the number of questions they attempt. The following chart gives a general guideline for how many questions you should be attempting.

Score	Questions
< 13	15/60
13 – 14	20/60
15 – 17	28/60
18 – 20	36/60
21 – 23	40/60
24 – 27	52/60
28 – 36	60/60

For example, a student with a current score of 19 should be attempting about 36 of the 60 questions on the test.

Since the math questions on the ACT tend to start out easier in the beginning of the section and get harder as you go, then attempting the first 36 questions is not a bad idea. However, it is okay to skip several questions and try a few that appear later on.

Note that although the questions tend to get harder as you go, it is not true that each question is harder than the previous question. For example, it is possible for question 25 to be easier than question 24, and in fact, question 25 can even be easier than question 20. But it is unlikely that question 50 would be easier than question 20.

If you are particularly strong in a certain subject area, then you may want to "seek out" questions from that topic even though they may be more difficult. For example, if you are very strong at number theory problems, but very weak at probability problems, then you may want to try every number theory problem no matter where it appears, and you may want to reduce the number of probability problems you attempt.

Remember that there is no guessing penalty on the ACT, so you should *not* leave any questions blank. This *does not* mean you should attempt every question. It means that if you are running out of time make sure you fill in answers for all the questions you did not have time to attempt.

For example, if you are currently scoring a 21, then it is possible you will only be attempting the first 40 questions or so. Therefore, when you are running out of time you should fill in answers for the last 20 problems. If you happen to get a chance to attempt some of them, you can always change your answer. But make sure those answers are filled in before the test ends!

10. Use your calculator wisely.

- Use a TI-84 or comparable calculator if possible when practicing and during the ACT.
- Make sure that your calculator has fresh batteries on test day.
- Make sure your calculator is in degree mode. If you are using a TI-84 (or equivalent) calculator press MODE and on the third line, make sure that DEGREE is highlighted. If it is not, scroll down and select it. If possible, do not alter this setting until you are finished taking your ACT.

Below are the most important things you should practice on your graphing calculator.

- Practice entering complicated computations in a single step.
- Know when to insert parentheses:
  - Around numerators of fractions
  - Around denominators of fractions
  - Around exponents
  - Whenever you actually see parentheses in the expression

Examples:

We will substitute a 5 in for x in each of the following examples.

Expression	Calculator computation
$\dfrac{7x+3}{2x-11}$	$(7*5+3)/(2*5-11)$
$(3x-8)^{2x-9}$	$(3*5-8)^\wedge(2*5-9)$

- Clear the screen before using it in a new problem. The big screen allows you to check over your computations easily.
- Press the **ANS** button (**2ND (-)**) to use your last answer in the next computation.
- Press **2ND ENTER** to bring up your last computation for editing. This is especially useful when you are plugging in answer choices, or guessing and checking.

- You can press **2ND ENTER** over and over again to cycle backwards through all the computations you have ever done.
- Know where the $\sqrt{}$, π, ^ and **LOG** buttons are so you can reach them quickly.
- Change a decimal to a fraction by pressing **MATH ENTER ENTER**.
- Press the **MATH** button - in the first menu that appears you can take cube roots and nth roots for any n. Scroll right to **NUM** and you have **lcm(** and **gcd(**. Scroll right to **PRB** and you have **nPr**, **nCr**, and **!** to compute permutations, combinations and factorials very quickly.
- Know how to use the **SIN**, **COS** and **TAN** buttons.

The following graphing tools can also be useful.

- Press the **Y=** button to enter a function, and then hit **ZOOM 6** to graph it in a standard window.
- Practice using the **WINDOW** button to adjust the viewing window of your graph.
- Practice using the **TRACE** button to move along the graph and look at some of the points plotted.
- Pressing **2ND TRACE** (which is really **CALC**) will bring up a menu of useful items. For example, selecting **ZERO** will tell you where the graph hits the x-axis, or equivalently where the function is zero. Selecting **MINIMUM** or **MAXIMUM** can find the vertex of a parabola. Selecting **INTERSECT** will find the point of intersection of 2 graphs.

14

PROBLEMS BY LEVEL AND TOPIC WITH FULLY EXPLAINED SOLUTIONS

Note: The quickest solution will always be marked with an asterisk (*).

LEVEL 1: NUMBER THEORY

1. $|5(-4) + 3(5)| = ?$

   A. -35
   B. -5
   C. 5
   D. 35
   E. 36

Recall that $|a|$ means the **absolute value** of a. It takes whatever number is between the two lines and makes it nonnegative. Here are a few examples: $|3| = 3$, $|-5| = 5$, $|0| = 0$.

* Quick solution: Simply type the following into your calculator:

$$5(-4) + 3(5)$$

The output is -5. But we want the **absolute value** of this number. So the answer to the question is 5, choice **C**.

Remarks: (1) You can use the absolute value function on your TI-84 calculator if you like. This can be found under the MATH menu. You would then type the following:

$$\text{abs}(5(-4) + 3(5)$$

The output will be 5, choice **C**.

(2) Note that I left off the rightmost parenthesis in the computation in Remark (1) above. There is no need to close parentheses at the end of an expression. Your calculator will do it automatically.

Solution by hand: $|5(-4) + 3(5)| = |-20 + 15| = |-5| = 5$, choice **C**.

Note: In the hand solution, all multiplication was done first, followed by the addition. Finally, the absolute value was taken at the end.

15

Order of Operations: Here is a quick review of order of operations.

PEMDAS	
P	Parentheses
E	Exponentiation
M	Multiplication
D	Division
A	Addition
S	Subtraction

Note that multiplication and division have the same priority, and addition and subtraction have the same priority.

2. The second term of an arithmetic sequence is 15 and the third term is 10. What is the first term?

   F. -15

   G. -10

   H. $\frac{1}{15}$

   J. 10

   K. 20

* Quick solution: Moving backwards, to get from the third term to the second term we add 5. Therefore, we add 5 more to get to the first term. So the first term is $15 + 5 = 20$, choice **K**.

Remark: In an arithmetic sequence, you always add (or subtract) the same number to get from one term to the next. This can be done by moving forwards or backwards through the sequence.

For the advanced student:

An **arithmetic sequence** is a sequence of numbers such that the difference **d** between consecutive terms is constant. The number **d** is called the **common difference** of the arithmetic sequence.

Here is an example of an arithmetic sequence: 20, 15, 10, 5, 0, -5, -10,... In this example the common difference is $d = 15 - 20 = -5$.

Note that this is the same arithmetic sequence given in the above ACT question.

Arithmetic sequence formula: $a_n = a_1 + (n - 1)d$

In the above formula, a_n is the nth term of the sequence. For example, a_1 is the first term of the sequence.

Note: In the arithmetic sequence 20, 15, 10, 5, 0, –5, –10,… we have that $a_1 = 20$ and $d = -5$. Therefore

$$a_n = 20 + (n-1)(-5) = 20 - 5n + 5 = 25 - 5n.$$

It follows that $a_1 = 25 - 5(1) = 25 - 5 = 20$, choice **K**.

Linear equations and arithmetic sequences: Questions about arithmetic sequences can easily be thought of as questions about lines and linear equations. We can identify terms of the sequence with points on a line where the x-coordinate is the term number and the y-coordinate is the term itself.

Remark: In the question above, since the second term of the sequence is 15, we can identify this term with the point (2,15). Since the third term of the sequence is 10, we can identify this with the point (3,10). Note that the common difference d is just the slope of the line that passes through these two points, i.e. $d = \frac{10-15}{3-2} = -5$.

3. Joseph bought a tie for 60% of its original price of $14.50 and a shirt for $\frac{2}{5}$ of the original price of $45.00. Ignoring sales tax, what is the total amount of these purchases?

    A. $21.00
    B. $25.00
    C. $26.00
    D. $26.70
    E. $40.50

* Quick calculator computation:

$$.60*14.50 + 2/5*45 = 26.70.$$

So the answer is choice **D**.

Note: (1) To change a percent to a decimal, divide by 100, or equivalently move the decimal point two places to the left (adding zeros if necessary). Note that the number 60 has an "invisible" decimal point after the 0 (so that 60 = 60.). Moving the decimal to the left two places gives us .60.

(2) The word "of" always translates to multiplication. So 60% of 14.50 is the same as .6(14.50), and $\frac{2}{5}$ of 45 is the same as $\frac{2}{5} \cdot 45$.

4. Dana needs $5\frac{1}{12}$ ounces of a chemical for an experiment. She has $3\frac{1}{4}$ ounces of the chemical. How many more ounces does she need?

    F. $1\frac{5}{12}$

    G. $1\frac{5}{6}$

    H. $2\frac{1}{6}$

    J. $2\frac{5}{6}$

    K. $2\frac{11}{12}$

* Quick calculator computation: We type the following into our TI-84 calculator: $5 + 1 / 12 - (3 + 1 / 4)$.

The output is 1.833333333. So it looks like the answer is choice G. To be safe, let's type in our calculator $1 + 5 / 6$. The output is also 1.833333333.

So the answer is choice **G**.

5. A pack of 50 balloons is priced at $3.50 now. If the balloons go on sale for 30% off the current price, what will be the sale price of the pack?

    A. $0.45
    B. $1.75
    C. $2.00
    D. $2.45
    E. $2.50

* Quick calculator computation:

$$.7 * 3.50 = 2.45$$

So the answer is choice **D**.

Remark: (1) "30% off" is the same as "70% of." So we are taking 70% of 3.50. As in problem 3 above, we change 70% to a decimal by moving the "invisible" decimal point to the left 2 places to get .7.

(2) We can also take 30 percent off of 3.50 by taking 30 percent **of** 3.50 and then subtracting this from 3.50.

So 30 percent of 3.50 is .3(3.50) = 1.05. Therefore 30 percent off of 3.50 is 3.50 – 1.05 = 2.45, choice **D**.

6.   Which of the following lists all the positive factors of 27 ?

    F.   1, 27
    G. 3, 9
    H. 3, 9, 27
    J.   27, 54, 81
    K. 1, 3, 9, 27

* Since (3)(9) = 27, 3 and 9 are both factors of 27. Also, (1)(27) = 27. So 1 and 27 are also factors of 27. So the answer is choice **K**.

Remark: It is easy to check whether one integer is a factor of another integer with our calculator. For example, if we divide 27 by 3 in our calculator we get 9. Since 9 is an integer, 3 is a factor of 27.

Definitions: The **integers** are the counting numbers together with their negatives.

$$\{\ldots,-4, -3, -2, -1, 0, 1, 2, 3, 4,\ldots\}$$

The **positive integers** consist of the positive numbers from that set.

$$\{1, 2, 3, 4,\ldots\}$$

An integer d is a **factor** of another integer n if there is an integer k such that $n = dk$. For example, 3 is a factor of 27 because 27 = (3)(9).

7.   What is the largest integer less than $\sqrt{73}$?

    A. 3
    B. 5
    C. 7
    D. 8
    E. 9

Solution by plugging in answer choices: Since the word "largest" appears in the problem, let's start with the largest answer choice, choice E. We have $9^2 = 81$. This is too big. So let's try choice D. Since $8^2 = 64$ and 64 < 73, we see that the answer is choice **D**.

* **Quick solution:** If we take the square root of 73 in our calculator we get approximately 8.544. The largest integer less than this is 8, choice **D**.

8. Philip earns $9.00 per hour for up to 40 hours of work in a week. For each hour over 40 hours of work in a week, Philip earns twice his regular pay. How much does Philip earn for a week in which he works 43 hours?

   F. $387.00
   G. $400.50
   H. $414.00
   J. $472.50
   K. $512.00

* Quick calculator computation:

$$9(40) + 18(3) = 414.$$

So Philip earns $414.00, choice **H.**

LEVEL 1: ALGEBRA AND FUNCTIONS

9. If $a = -5$, what is the value of $\frac{a^2 - 4}{a+2}$?

   A. -7
   B. -4
   C. 4
   D. $9\frac{2}{3}$
   E. 12

We have $a^2 - 4 = (-5)^2 - 4 = 25 - 4 = 21$. Also $a + 2 = -5 + 2 = -3$. Finally, we divide $21 / (-3) = -7$, choice **A.**

Remarks: (1) If you are doing these computations on your calculator, make sure that -5 is put in parentheses before squaring: $-5 \wedge 2$ will give an output of -25 which is not correct.

(2) This can be done with a single calculator computation as follows:

$$((-5) \wedge 2 - 4) / (-5 + 2)$$

Note that the whole numerator is inside parentheses and the whole denominator is inside parentheses.

10. $x^2 - 73x + 27 - 46x^2 + 75x$ is equivalent to:

    F. $-29x^2$

    G. $-29x^6$

    H. $-45x^4 + 2x^2 + 27$

    J. $-45x^2 + 2x + 27$

    K. $-44x^2 + 2x + 27$

Solution by picking a number: Let's choose a value for x, say $x = 2$. Then

$$x^2 - 73x + 27 - 46x^2 + 75x = 2^2 - 73(2) + 27 - 46(2)^2 + 75(2) = \mathbf{-149}$$

Put a nice big dark circle around **−149** so you can find it easier later. We now substitute 2 for x into each answer choice:

    F. $-29(2)^2 = -116$

    G. $-29(2)^6 = -1856$

    H. $-45(2)^4 + 2(2)^2 + 27 = -685$

    J. $-45x^2 + 2x + 27 = -149$

    K. $-44x^2 + 2x + 27 = -145$

Since F, G, H, and K each came out incorrect, the answer is choice **J**.

Important note: J is **not** the correct answer simply because it is equal to -149. It is correct because all four of the other choices are **not** -149. **You absolutely must check all five choices!**

* Algebraic solution:

$$x^2 - 73x + 27 - 46x^2 + 75x = x^2 - 46x^2 - 73x + 75x + 27$$
$$= (1 - 46)x^2 + (-73 + 75)x + 27 = -45x^2 + 2x + 27$$

This is choice **J**.

11. If $4b - 5 = 17$, then $b =$

    A.   4.0

    B.   5.5

    C.   10.0

    D.   17.5

    E.   22.0

Solution by plugging in answer choices: Let's start with choice C and guess that $b = 10$. Then $4b - 5 = 4(10) - 5 = 40 - 5 = 35$. This is too big. So we can eliminate choices C, D, and E.

Let's try choice B next. So we are guessing that $b = 5.5$. We then have that $4b - 5 = 4(5.5) - 5 = 22 - 5 = 17$. This is correct. So the answer is choice **B.**

 * **Algebraic solution:** We add 5 to each side of the equation $4b - 5 = 17$ to get $4b = 22$. We then divide each side of this equation by 4 to get that $b = 5.5$, choice **B.**

12. Which of the following is an equivalent simplified expression for $3(5x + 8) - 4(3x - 2)$?

    F. $x + 16$
    G. $3x + 16$
    H. $3x + 32$
    J. $7x + 2$
    K. $7x + 3$

Solution by picking a number: Let's choose a value for x, say $x = 2$. Then

$$3(5x + 8) - 4(3x - 2) = 3(5 \cdot 2 + 8) - 4(3 \cdot 2 - 2) = 3(10 + 8) - 4(6 - 2)$$
$$= 3(18) - 4(4) = 54 - 16 = \mathbf{38}.$$

Put a nice big dark circle around **38** so you can find it easier later. We now substitute 2 for x into each answer choice.

    F. 18
    G. 22
    H. 38
    J. 16
    K. 17

Since F, G, J, and K each came out incorrect, the answer is choice **H.**

Important note: H is **not** the correct answer simply because it is equal to 38. It is correct because all four of the other choices are **not** 38. **You absolutely must check all five choices!**

Remark: The computation $3(5 \cdot 2 + 8) - 4(3 \cdot 2 - 2)$ can be done in a single step with your calculator. Simply input the following.

$$3(5 * 2 + 8) - 4(3 * 2 - 2) \text{ ENTER}$$

* Algebraic solution:

$$3(5x + 8) - 4(3x - 2) = 15x + 24 - 12x + 8 = 3x + 32$$

This is choice **H**.

Note: Make sure you are using the distributive property correctly here. For example, $3(5x + 8) = 15x + 24$. A common mistake would be to write $3(5x + 8) = 15x + 8$.

13. A Celsius temperature C can be approximated by subtracting 32 from the Fahrenheit temperature F and then multiplying by $\frac{1}{2}$. Which of the following expresses this approximation method? (Note: The symbol $\approx$ means "is approximately equal to.")

   A. $C \approx \frac{1}{2}(F - 32)$

   B. $C \approx \frac{1}{2}F - 32$

   C. $C \approx 2(F - 32)$

   D. $C \approx 2F - 32$

   E. $C \approx \sqrt{F} - 32$

* When we subtract 32 from F we get $F - 32$. Multiplying this expression by $\frac{1}{2}$ yields $\frac{1}{2}(F - 32)$. This is choice **A**.

Caution: A common mistake would be to multiply only the first term by $\frac{1}{2}$ to get $\frac{1}{2}F - 32$. This is wrong. Whenever you are doing anything to an expression always keep it in parentheses as was done in the solution above.

14. Which of the following expressions is equivalent to $5a + 10b + 15c$?

   F. $5(a + 2b + 3c)$

   G. $5(a + 2b + 15c)$

   H. $5(a + 10b + 15c)$

   J. $5(a + 2b) + 3c$

   K. $30(a + b + c)$

Solution by picking numbers: Let's choose values for a, b, and c, say $a = 2$, $b = 3$, $c = 4$. Then

$$5a + 10b + 15c = 5(2) + 10(3) + 15(4) = 10 + 30 + 60 = \mathbf{100}.$$

23

Put a nice big dark circle around **100** so you can find it easier later. We now substitute $a = 2$, $b = 3$, $c = 4$ into each answer choice:

F. $5(2 + 2 \cdot 3 + 3 \cdot 4) \quad = 100$

G. $5(2 + 2 \cdot 3 + 15 \cdot 4) = 340$

H. $5(2 + 10 \cdot 3 + 15 \cdot 4) = 460$

J. $5(2 + 2 \cdot 3) + 3 \cdot 4 \quad = \ 52$

K. $30(2 + 3 + 4) \quad = 270$

Since G, H, J, and K each came out incorrect, the answer is choice **F**.

Important note: F is **not** the correct answer simply because it is equal to 100. It is correct because all four of the other choices are **not** 100. **You absolutely must check all five choices!**

Remark: All of the above computations can be done in a single step with your calculator.

* Algebraic solution: We simply factor out a 5 to get

$$5a + 10b + 15c = 5(a + 2b + 3c)$$

This is choice **F**.

Remark: (1) If you have trouble seeing why the right hand side is the same as what we started with on the left, try working backwards and multiplying instead of factoring. In other words, we have

$$5(a + 2b + 3c) = 5a + 10b + 15c$$

Note how the distributive property is being used here. Each term in parentheses is multiplied by the 5!

(2) You can also start with the answer choices and do each multiplication (as we did in Remark (1)) until you get $5a + 10b + 15c$.

15. Which of the following is a value for z that solves the equation $|z - 4| = 9$?

    A. -13

    B. $- 5$

    C. $\dfrac{9}{4}$

    D. 5

    E. 36

24

Solution by plugging in answer choices: Normally I would start with choice C, but in this case there are much simpler numbers to plug in, so let's start with choice B and guess that $z = -5$. Then we have

$$|z - 4| = |-5 - 4| = |-9| = 9.$$

This is correct. So the answer is choice **B.**

* **Algebraic solution:** The absolute value equation $|z - 4| = 9$ is equivalent to the **two** equations

$$z - 4 = 9 \qquad\qquad z - 4 = -9$$

Adding 4 to each side of each of these equations yields $z = 9 + 4 = 13$ or $z = -9 + 4 = -5$. Only the latter answer is an answer choice. So the answer is choice **B.**

16. If $c > 1$, then which of the following has the <u>least</u> value?

    F. $\sqrt{c}$
    G. $\sqrt{2c}$
    H. $\sqrt{c^2}$
    J. $c\sqrt{c}$
    K. c^2

Solution by picking a number: Let's choose a value for c that is greater than 1, say $c = 2$, and plug this value into each answer choice (using our calculator to get an approximate answer).

    F. $\sqrt{2}$ ≈ 1.414
    G. $\sqrt{2 \cdot 2}$ $= 2$
    H. $\sqrt{2^2}$ $= 2$
    J. $2\sqrt{2}$ ≈ 2.828
    K. 2^2 $= 4$

We see that the least value appears in choice **F.**

LEVEL 1: GEOMETRY

17. In $\triangle PQR$, the sum of the measures of $\angle P$ and $\angle Q$ is 59°. What is the measure of $\angle R$?

    A. 31°
    B. 59°
    C. 118°
    D. 121°
    E. 131°

Solution by plugging in answer choices: First note that the three angle measures in a triangle add up to 180°.

Let's start with choice C and guess that the measure of angle R is 118°. We have 59 + 118 = 177. This is a bit too small. So we can eliminate choices A, B, and C.

Let's try choice D next and guess that the measure of angle R is 121°. We have 59 + 121 = 180. This is correct. So the answer is **D.**

* **Quick solution:** Since the three angle measures in a triangle add up to 180 degrees, the measure of angle R is 180 − 59 = 121°, choice **D.**

Note: 59° is the <u>sum</u> of two angle measures. So adding this to the third angle measure gives 180°.

Solution by drawing a picture and picking numbers: Let's draw a picture.

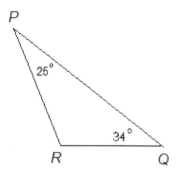

In the above picture I chose angle P and angle Q to have measures that add to 59 degrees. It follows that the measure of angle R, in degrees is 180 − 25 − 34 = 121, choice **D.**

Definition: A **triangle** is a two-dimensional geometric figure with three sides and three angles. The sum of the degree measures of all three angles of a triangle is 180°.

18. What is the perimeter, in centimeters, of a rectangle with length 15 in and width 3 in?

    F. 18
    G. 21
    H. 36
    J. 45
    K. 90

* Quick solution: $P = 2\ell + 2w = 2(15) + 2(3) = 30 + 6 = 36$, choice **H.**

Here is a picture for extra clarification.

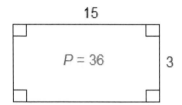

Definitions: A **quadrilateral** is a two-dimensional geometric figure with four sides and four angles. The sum of the degree measures of all four angles of a quadrilateral is 360.

A **rectangle** is a quadrilateral in which each angle is a right angle. That is, each angle measures 90°.

The **perimeter** of a rectangle is $P = 2\ell + 2w$.

19. In parallelogram *PQRS*, which of the following must be true about the measures of ∠*PQR* and ∠*QRS* ?

    A. each are 90°
    B. each are less than 90°
    C. each are greater than 90°
    D. they add up to 90°
    E. they add up to 180°

Solution by drawing a picture and process of elimination: Let's draw a picture.

27

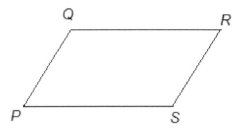

Note that the measure of angle *PQR* is greater than 90 degrees, and the measure of angle *QRS* is less than 90 degrees. So we can eliminate choices A, B, and C.

Since the measure of angle *PQR* by itself is greater than 90 degrees, we can eliminate choice D. Therefore, the answer is choice **E.**

Note: If you have trouble eliminating choices from the general picture above, try choosing specific angle measures. Here is an example:

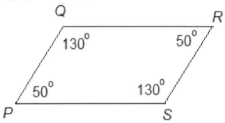

Observe that we had to choose angle measures that sum to 360 (since *PQRS* is a quadrilateral). We also had to make sure that opposite angles were congruent (since *PQRS* is a parallelogram).

* Quick solution: If you happen to recall that adjacent angles of a parallelogram are **supplementary** (have measures which add to 180 degrees), you could choose choice **E** immediately (or after drawing the first picture above, if necessary).

Facts about parallelograms:

(1) opposite sides are congruent
(2) opposite sides are parallel
(3) opposite angles are congruent
(4) the diagonals bisect each other

20. A point at (−5,6) in the standard (x, y) coordinate plane is shifted up 4 units and right 8 units. What are the coordinates of the new point?

F. (−1,14)
G. (−13,10)
H. (−13, 2)
J. (3, 2)
K. (3,10)

Solution by drawing a picture: Let's a draw a picture of this situation.

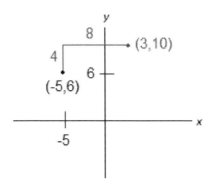

Note that to plot the point (−5,6), from the origin (0,0) we move left 5 and up 6. From there we move up 4 and right 8 to get to the point (3,10), choice **K**.

Note: We are adding 4 to the y-coordinate of the point and 8 to the x-coordinate of the point: $6 + 4 = 10$ and $−5 + 8 = 3$.

21. The interior dimensions of a rectangular box are 5 inches by 4 inches by 3 inches. What is the volume, in cubic inches, of the interior of the box?

A. 12
B. 60
C. 90
D. 120
E. 124

The volume of the box, in inches, is simply $V = \ell wh = (5)(4)(3) = 60$, choice **B**.

22. Given right triangle ΔPQR below, what is the length of $\overline{PQ}$?

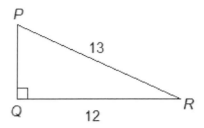

F. $\sqrt{2}$
G. $\sqrt{5}$
H. 5
J. 7
K. 11

* Solution using Pythagorean triples: We use the Pythagorean triple 5, 12, 13 to see that $PQ = 5$, choice **H**.

Note: The most common Pythagorean triples are 3,4,5 and 5, 12, 13. Two others that may come up are 8, 15, 17 and 7, 24, 25.

Solution by the Pythagorean Theorem: By the Pythagorean Theorem, we have $13^2 = (PQ)^2 + 12^2$. So $169 = (PQ)^2 + 144$. Subtracting 144 from each side of this equation yields $25 = (PQ)^2$, or $PQ = 5$, choice **H**.

Remarks: (1) The Pythagorean Theorem says that if a right triangle has legs of length a and b, and a hypotenuse of length c, then $c^2 = a^2 + b^2$.

(2) Be careful in this problem: the length of the hypotenuse is 13. So we replace c by 13 in the Pythagorean Theorem

(3) The equation $x^2 = 25$ would normally have two solutions: $x = 5$ and $x = -5$. But the length of a side of a triangle cannot be negative, so we reject -5.

23. On a real number line, point X is at -3.25 and is 6.75 units from point Y. What are the possible locations of point Y on the real number line?

   A. -10 and -3.5
   B. -10 and 3.5
   C. -10 and 10
   D. 10 and -3.5
   E. 10 and 3.5

30

* We compute –3.25 + 6.75 = 3.5 and –3.25 – 6.75 = –10, choice **B.**

Here is a picture illustrating this problem.

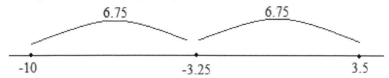

24. In the figure below, adjacent sides meet at right angles and the lengths given are in inches. What is the perimeter of the figure, in inches?

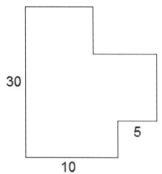

    F.   45
    G.   60
    H.   90
    J.   120
    K. 450

* **Solution by moving the sides of the figure around:** Recall that to compute the perimeter of the figure we need to add up the lengths of all 8 line segments in the figure. We "move" the two smaller vertical segments to the right, and each of the smaller horizontal segments up or down as shown below.

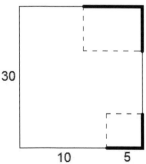

31

Note that the "bold" length is equal to the "dashed" length. We get a rectangle with length 30 and width 15. Thus, the perimeter is

$$(2)(30) + (2)(15) = 60 + 30 = 90.$$

This is choice **H**.

Warning: Although lengths remain unchanged by moving line segments around, areas will be changed. This method should **not** be used in problems involving areas.

LEVEL 1: PROBABILITY AND STATISTICS

25. Jason has taken 4 tests in his math class, with grades of 73, 86, 64, and 97. In order to maintain this exact average, what *must* be Jason's grade on his 5th math test?

   A. 60
   B. 70
   C. 80
   D. 90
   E. 95

Jason's average on the 4 tests is $\frac{73+86+64+97}{4} = 80$. In order to maintain this exact average, Jason must receive an 80 on his 5th test, choice **C**.

Notes: (1) The **average (arithmetic mean)** of a list of numbers is the sum of the numbers in the list divided by the quantity of the numbers in the list.

$$\textbf{Average} = \frac{\textbf{Sum}}{\textbf{Number}}$$

(2) When we add a new number to a list of numbers, the average will remain the same precisely when the new number is equal to the old average.

If the new number is greater than the old average, the average will increase, and if the new number is less than the old average, the average will decrease.

26. A menu lists 2 appetizers, 5 meals, 4 drinks, and 3 desserts. A dinner consists of 1 of each of these 4 items. How many different dinners are possible from this menu?

F. 2
G. 4
H. 14
J. 72
K. 120

* Solution using the counting principle: $(2)(5)(4)(3) = 120$, choice **K**.

Remark: The **counting principle** says that if one event is followed by a second independent event, the number of possibilities is multiplied.

More generally, if $E_1, E_2, \ldots, E_n$ are n independent events with $m_1, m_2, \ldots, m_n$ possibilities, respectively, then event E_1 followed by event E_2, followed by event E_3, ..., followed by event E_n has $m_1 \cdot m_2 \cdots m_n$ possibilities.

In this question there are 4 events: "choosing an appetizer," "choosing a meal," "choosing a drink," and "choosing a dessert."

27. In an urn with 60 marbles, 20% of the marbles are black. If you randomly choose a marble from the urn, what is the probability that the marble chosen is <u>not</u> one of the black marbles?

A. $\dfrac{1}{2}$

B. $\dfrac{1}{5}$

C. $\dfrac{2}{5}$

D. $\dfrac{3}{5}$

E. $\dfrac{4}{5}$

* 20% of 60 is $(0.2)(60) = 12$. So 12 of the marbles are black. Therefore, $60 - 12 = 48$ of the marbles are not black. So the requested probability is $\dfrac{48}{60} = \dfrac{4}{5}$, choice **E**.

Remarks: (1) To compute a simple probability where all outcomes are equally likely, divide the number of "successes" by the total number of outcomes.

33

In this problem, the total is 60 marbles, and 48 of them are "successes."

(2) You can quickly reduce a fraction in your TI-84 calculator by performing the division and then pressing MATH ENTER ENTER.

Alternate solution: As in the first solution we see that 12 of the marbles are black. So the probability of choosing a black marble is $\frac{12}{60} = \frac{1}{5}$. The probability of choosing a marble that is not black is $1 - \frac{1}{5} = \frac{4}{5}$, choice **E**.

Note: If P(E) is the probability of event E occurring, then the probability of event E <u>not</u> occurring is $1 - $ P(E).

28. A data set contains 6 elements and has a mean of 5. Five of the elements are 2, 4, 6, 8, and 10. Which of the following is the sixth element?

    F. 0
    G. 1
    H. 2
    J. 3
    K. 4

* Solution by changing the average to a sum: We change the average (or mean) to a sum using the formula

$$\text{Sum} = \text{Average} \cdot \text{Number}$$

Here we are averaging 6 elements. Thus, the **Number** is 6. The **Average** is given to be 5. Therefore, the **Sum** of the 6 numbers is $5 \cdot 6 = 30$. The sixth element is therefore

$$30 - 2 - 4 - 6 - 8 - 10 = 0$$

This is choice **F**.

29. If the probability that it will be sunny tomorrow is 0.4, what is the probability that it will <u>not</u> be sunny tomorrow?

    A. 1.4
    B. 1.0
    C. 0.6
    D. 0.1
    E. 0.0

* The requested probability is $1 - 0.4 = 0.6$, choice **C**.

See the note at the end of question 27 for further explanation.

30. The average of 8 numbers is 7.3. If each of the numbers is decreased by 6, what is the average of the 8 new numbers?

    F. 0.0
    G. 0.3
    H. 1.3
    J. 2.3
    K. 7.3

Solution by changing the average to a sum: We change the average (or mean) to a sum using the formula

$$\text{Sum} = \text{Average} \cdot \text{Number}$$

Here we are averaging 8 numbers. Thus the **Number** is 8. The **Average** is given to be 7.3. Therefore, the **Sum** of the 8 numbers is $7.3 \cdot 8 = 58.4$.

Since each number is decreased by 6, the sum is decreased by $6 \cdot 8 = 48$. So the sum of the 8 new numbers is $58.4 - 48 = 10.4$, and the average of the 8 new numbers is $\frac{10.4}{8} = 1.3$, choice **H.**

* Quick solution: Since each number is decreased by the same amount, the average is decreased by this amount as well. So the average of the 8 new numbers is $7.3 - 6 = 1.3$, choice **H.**

31. Markus has 4 white hats and 5 black hats in his closet. If he randomly takes 1 of these 9 hats from his closet, what is the probability that the hat that Markus takes is black?

    A. $\dfrac{1}{9}$

    B. $\dfrac{1}{5}$

    C. $\dfrac{4}{9}$

    D. $\dfrac{5}{9}$

    E. $\dfrac{5}{4}$

* Recall from the first Remark in problem 27 that to compute a simple probability where all outcomes are equally likely, we divide the number of "successes" by the total number of outcomes.

In this problem, the total is 9 hats, and 5 of them are "successes." So the probability that the hat Markus takes is black is $\frac{5}{9}$, choice **D**.

32. Debra's grades on the first 3 tests in her math class were 86, 74, and 93. How many points must Debra receive on the fourth test to average exactly 85 points for these 4 tests (assume that all 4 tests are equally weighted)?

    F. 86
    G. 87
    H. 88
    J. 89
    K. 90

Solution by changing the average to a sum: We change the given average to a sum using the formula

$$\text{Sum} = \text{Average} \cdot \text{Number}$$

Here we are averaging 4 test grades. Thus, the **Number** is 4. The **Average** of the 4 test grades is given to be 85. Therefore, the **Sum** of the 4 test grades is $85 \cdot 4 = 340$. So the number of points Debra must receive on the fourth test is

$$340 - 86 - 74 - 93 = 87$$

This is choice **G**.

* Solution by "balancing" each side of the average: The desired average is 85. 86 is 1 more than this average, 74 is 11 less than this average, and 93 is 8 more than this average. So the "balancing factor" is

$$+1 - 11 + 8 = -2$$

We can bring the balance back to 0 with a +2. So the number of points Debra must receive on the fourth test is $85 + 2 = 87$, choice **G**.

Solution by plugging in answer choices: Let's start with choice H and guess that the number of points Debra must receive on the fourth test is 88. It follows that the average is $\frac{86+74+93+88}{4} = 85.25$. Since this is greater than 85, we can eliminate choices H, J, and K.

Let's try choice G next. So we are guessing the number of points Debra must receive on the fourth test is 87. It follows that the average is $\frac{86+74+93+87}{4} = 85$. This is correct. So the answer is choice **G**.

LEVEL 2: NUMBER THEORY

33. The expression $\dfrac{5+\frac{1}{5}}{2+\frac{1}{10}}$ is equal to:

   A. $\dfrac{7}{2}$

   B. $\dfrac{52}{21}$

   C. 3

   D. 7

   E. 9

* Calculator solution: We simply type the following into our calculator:

$$(5 + 1 / 5) / (2 + 1 / 10) \text{ ENTER MATH ENTER ENTER}$$

The output is 52/21. So the answer is choice **B**.

Hand solution (not recommended): Note that this is a **complex fraction**. Inside this complex fraction are the two **simple fractions** $\frac{1}{5}$ and $\frac{1}{10}$. The least common denominator of these simple fractions is 10.

So we multiply the numerator and denominator of the complex fraction by 10:

$$\frac{10\left(5+\frac{1}{5}\right)}{10\left(2+\frac{1}{10}\right)} = \frac{50+2}{20+1} = \frac{52}{21}$$

This is choice **B**.

Note: Make sure you distribute the 10 correctly when multiplying in the numerator and denominator of the fraction. For example, in the numerator, you need to multiply 5 by 10 **and** to multiply $\frac{1}{5}$ by 10.

34. What is the least common multiple of 100, 70, and 30?

   F. 210,000

   G. 2,100

   H. 210

   J. 180

   K. 60

Solution by plugging in answer choices: Since the word "least" appears in the problem, let's start with the smallest answer choice, choice K. Since 60 is smaller than 100 it cannot be a multiple of 100. So we can eliminate choice K.

180 and 210 are not multiples of 100 (because they do not end in two zeros). So we can eliminate choices J and H.

Now 2,100 does end in two zeros, and so it is divisible by 100. Also, we have $\frac{2,100}{70} = 30$ and $\frac{2,100}{30} = 70$. Since these are all integers, the answer is choice **G.**

* Calculator solution: We use the **lcm** feature on our graphing calculator (found under NUM after pressing the MATH button). Our calculator can only handle two numbers at a time. So compute **lcm**(100,70) = 700, and then **lcm**(700,30) = 2100, choice **G.**

Solution using prime factorizations:

Step 1: Find the prime factorization of each integer in the set.

$$100 = 2^2 \cdot 5^2$$
$$70 = 2 \cdot 5 \cdot 7$$
$$30 = 2 \cdot 3 \cdot 5$$

Step 2: Choose the highest power of each prime that appears in any of the factorizations and multiply them together:

$$2^2 \cdot 3 \cdot 5^2 \cdot 7 = 2100, \text{ choice } \textbf{G.}$$

Here is a quick lesson in **prime factorization**

The Fundamental Theorem of Arithmetic: Every integer greater than 1 can be written "uniquely" as a product of primes.

The word "uniquely" is written in quotes because prime factorizations are only unique if we agree to write the primes in increasing order.

For example, 30 can be written as $2 \cdot 3 \cdot 5$ or as $3 \cdot 2 \cdot 5$ (as well as several other ways). But these factorizations are the same except that we changed the order of the factors.

To make things simple we always agree to use the **canonical representation**. The word "canonical" is just a fancy name for "natural," and the most natural way to write a prime factorization is in increasing order of primes. So the canonical representation of 30 is $2 \cdot 3 \cdot 5$.

As another example, the canonical representation of 100 is $2 \cdot 2 \cdot 5 \cdot 5$. We can tidy this up a bit by rewriting $2 \cdot 2$ as 2^2 and $5 \cdot 5$ as 5^2. So the canonical representation of 100 is $2^2 \cdot 5^2$.

If you are new to factoring, you may find it helpful to draw a factor tree. For example, here is a factor tree for 100:

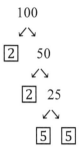

To draw this tree we started by writing 100 as the product $2 \cdot 50$. We put a box around 2 because 2 is prime, and does not need to be factored anymore. We then proceeded to factor 50 as $2 \cdot 25$. We put a box around 2, again because 2 is prime. Finally, we factor 25 as $5 \cdot 5$. We put a box around each 5 because 5 is prime. We now see that we are done, and the prime factorization can be found by multiplying all of the boxed numbers together. Remember that we will usually want the canonical representation, so write the final product in increasing order of primes.

By the Fundamental Theorem of Arithmetic above it does not matter how we factor the number – we will always get the same canonical form. For example, here is a different factor tree for 100:

$$100$$
$$4 \qquad 25$$
$$\boxed{2} \ \boxed{2} \quad \boxed{5} \ \boxed{5}$$

Definitions: (1) A **prime number** is a positive integer that has **exactly** two factors (1 and itself). Here is a list of the first few primes:

$$2, 3, 5, 7, 11, 13, 17, 19, 23,\ldots$$

Note that 1 is **not** prime. It only has one factor!

(2) The **least common multiple (lcm)** of a set of positive integers is the smallest positive integer that is divisible by each integer in the set.

35. What is the value of $|-5| - |11 - 29|$?

  A. -45
  B. -23
  C. -13
  D. 13
  E. 23

Recall that $|a|$ means the **absolute value** of a. It takes whatever number is between the two lines and makes it nonnegative. For some examples, we have $|3| = 3, |-5| = 5, |0| = 0$.

* Calculator solution: Simply type the following into your calculator:

$$\text{abs}(-5) - \text{abs}(11 - 29)$$

The output is -13, choice **C**.

Remarks: (1) The absolute value function on your TI-84 calculator can be found under the MATH menu.

Hand solution: $|-5| - |11 - 29| = 5 - |-18| = 5 - 18 = -13$, choice **C**.

36. How many minutes would it take a car to travel 42 miles at a constant speed of 56 miles per hour?

  F. 90
  G. 80
  H. 45
  J. 40
  K. 30

* Using $d = r \cdot t$ (distance = rate $\cdot$ time), we have

$$42 = 56t$$

$$t = \frac{42}{56} = \frac{3}{4} \text{ hours}$$

But $\frac{3}{4}$ of an hour is 45 minutes, choice **H**.

37. A small theatre has 12 rows of seats. The front row has 22 seats and each succeeding row has 1 less seat than the row in front of it. How many seats will be in the back row?

  A. 9
  B. 10
  C. 11
  D. 12
  E. 13

40

Solution by listing: Let's simply form a list of numbers, starting with 22, and decreasing by 1 for each new term, until we have written out 12 terms:

$$22, 21, 20, 19, 18, 17, 16, 15, 14, 13, 12, 11$$

Note that we have written out 12 numbers and the last number is 11, choice **C**.

* Solution using an arithmetic sequence: Recall, from problem 2, the arithmetic sequence formula: $a_n = a_1 + (n - 1)d$. In this question the first term is $a_1 = 22$ and the common difference is $d = -1$. We want the 12th term which is $a_{12} = 22 + (12 - 1)(-1) = 22 - 11 = 11$, choice **C**.

38. If you add up 5 consecutive odd integers that are each less than 32, what is the largest possible sum?

    F.   25
    G. 87
    H. 125
    J.   135
    K. 145

* We have $31 + 29 + 27 + 25 + 23 = 135$, choice **J**.

Remarks: (1) We get the largest sum by making the integers as large as possible. The largest odd integer less than 32 is 31.

Consecutive odd integers are odd integers that follow each other in order. The difference between consecutive odd integers is 2. Here are two examples.

        1, 3, 5   these are three consecutive odd integers
  -5, -3, -1, 1, 3   these are five consecutive odd integers

In general, if x is an odd integer, then $x, x + 2, x + 4, x + 6,...$ are consecutive odd integers.

39. For all positive integers k, what is the greatest common factor of the 2 numbers $150k$ and $700k$?

    A. 10
    B. 50
    C. k
    D. $10k$
    E. $50k$

41

Solution by plugging in answer choices: Since the word "greatest" appears in the problem, let's start with the largest answer choice, choice E. Now $\frac{150k}{50k} = 30$ and $\frac{700k}{50k} = 14$. Since these are both integers, the answer is choice **E.**

* Calculator solution: Compute **gcd**$(150,700) = 50$. It follows that $\gcd(150k,700k) = 50k$, choice **E.**

Solution using prime factorizations: $150 = 2 \cdot 3 \cdot 5^2$ and $700 = 2^2 \cdot 5^2 \cdot 7$. So $\gcd(150,700) = 2 \cdot 5^2 = 50$. It follows that $\gcd(150k,700k) = 50k$, choice **E.**

Notes: (1) See the end of the solution to problem 34 for a lesson in finding prime factorizations.

(2) gcd stands for "greatest common divisor." Divisor is just another word for factor.

(3) To find the gcd of the given numbers, we choose the lowest power of each prime that appears in **both** of the factorizations and multiply them together (for example, 3 appears in the factorization of 150, but **not** 700, so it does not contribute to the gcd).

40. What is the correct ordering of 2π, 6, and $\frac{13}{2}$ from least to greatest?

   F. $6 < 2\pi < \frac{13}{2}$

   G. $2\pi < 6 < \frac{13}{2}$

   H. $2\pi < \frac{13}{2} < 6$

   J. $\frac{13}{2} < 6 < 2\pi$

   K. $6 < \frac{13}{2} < 2\pi$

Solution by converting to decimals: We rewrite each number as a decimal (or decimal approximation) as follows:

$$2\pi \approx 6.28 \qquad \frac{13}{2} = 6.5$$

Now let's line them up in order: $6 < 6.28 < 6.5$

So we have $6 < 2\pi < \frac{13}{2}$, choice **F.**

Note: We can compare two decimals by looking at the first position where they disagree. For example, 6.28 is less than 6.5 because 2 is less than 5. If a digit is missing, there is a hidden 0 there. Thus 6 is less than 6.28 because 6 is the same as 6.0 and 0 is less than 2.

LEVEL 2: ALGEBRA AND FUNCTIONS

41. A function $g(x)$ is defined as $g(x) = -5x^2$. What is $g(-2)$?

    A. −20
    B.   20
    C.   50
    D. −100
    E.   100

* $g(-2) = -5(-2)^2 = -5(4) = -20$, choice **A**.

Notes: (1) The variable x is a placeholder. We evaluate the function g at a specific value by substituting that value in for x. In this question we replaced x by −2.

(2) The exponentiation was done first, followed by the multiplication. See the end of the solution to problem 1 for more information on order of operations.

(3) To square a number means to multiply it by itself. So

$$(-2)^2 = (-2)(-2) = 4.$$

(4) We can do the whole computation in our calculator in one step. Simply type -5(-2)^2 ENTER. The output will be -20.

Make sure to use the minus sign and not the subtraction symbol. Otherwise the calculator will give an error.

42. Let a function of 2 variables be defined by $h(x,y) = x^2 + 3xy - (y - x)$, what is the value of $h(5,4)$?

    F.   76
    G.   84
    H.   85
    J.   86
    K.   94

* $h(5,4) = 5^2 + 3(5)(4) - (4 - 5) = 25 + 60 - (-1) = 85 + 1 = 86$, choice **J**.

Notes: (1) This is very similar to problem 41. Everywhere we see an x we replace it by 5 and everywhere we see a y we replace it by 4. Remember to follow the correct order of operations (see problem 1 for more details).

(2) We can do the whole computation in our calculator in one step. Simply type $5\hat{\ }2 + 3*5*4 - (4 - 5)$ ENTER. The output will be 86.

43. $5y^4 \cdot 11y^4$ is equivalent to

    F. $16y^8$
    G. $16y^{16}$
    H. $55y^4$
    J. $55y^8$
    K. $55y^{16}$

Solution by picking a number: Let's choose a value for y, say $y = 2$. Then

$$5y^4 \cdot 11y^4 = 5(2)^4 \cdot 11(2)^4 = 5(16) \cdot 11(16) = \mathbf{14{,}080}$$

Put a nice big dark circle around **14,080** so you can find it easier later. We now substitute 2 for y into each answer choice:

    F. $16(2)^8 = 4{,}096$
    G. $16(2)^{16} = 1{,}048{,}576$
    H. $55(2)^4 = 880$
    J. $55(2)^8 = 14{,}080$
    K. $55(2)^{16} = 3{,}604{,}480$

Since F, G, H, and K each came out incorrect, the answer is choice **J**.

Important note: J is **not** the correct answer simply because it is equal to 14,080. It is correct because all four of the other choices are **not** 14,080. **You absolutely must check all five choices!**

* Algebraic solution: $5y^4 \cdot 11y^4 = 5 \cdot 11 \cdot y^4 \cdot y^4 = 55y^{4+4} = 55y^8$, choice **J**.

Notes: (1) The set of real numbers is **commutative** and **associative** for multiplication. These two properties together allow us to rearrange factors in any order we like when we are multiplying several numbers together.

More precisely, the set of real numbers is commutative for multiplication because for any two real numbers a and b, we have $a \cdot b = b \cdot a$. Also the set of real numbers is associative for multiplication because for any three real numbers a, b, and c, we have $a \cdot (b \cdot c) = (a \cdot b) \cdot c$.

(2) In this problem we used the law of exponents illustrated in the third row in the table below.

Laws of Exponents: For those students that have forgotten, here is a brief review of the basic laws of exponents.

Law	Example
$x^0 = 1$	$3^0 = 1$
$x^1 = x$	$9^1 = 9$
$x^a x^b = x^{a+b}$	$x^3 x^5 = x^8$
$x^a / x^b = x^{a-b}$	$x^{11} / x^4 = x^7$
$(x^a)^b = x^{ab}$	$(x^5)^3 = x^{15}$
$(xy)^a = x^a y^a$	$(xy)^4 = x^4 y^4$
$(x/y)^a = x^a / y^a$	$(x/y)^6 = x^6 / y^6$

44. What polynomial must be added to $x^2 + 3x - 5$ so that the sum is $5x^2 - 8$?

    F. $4x^2 - 5x + 6$
    G. $4x^2 - 3x - 3$
    H. $5x^2 - 3x - 3$
    J. $5x^2 - 3x + 6$
    K. $6x^2 + 3x - 13$

Solution by picking a number: Let's choose a value for x, say $x = 2$. Then we have $x^2 + 3x - 5 = (2)^2 + 3(2) - 5 = 4 + 6 - 5 = 5$ and we also have $5x^2 - 8 = 5(2)^2 - 8 = 5(4) - 8 = 20 - 8 = 12$. So we must add **7** to get from 5 to 12 (indeed, $12 - 5 = 7$).

Put a nice big dark circle around **7** so you can find it easier later. We now substitute 2 for x into each answer choice:

    F. $4(2)^2 - 5(2) + 6 = 12$
    G. $4(2)^2 - 3(2) - 3 = 7$
    H. $5(2)^2 - 3(2) - 3 = 11$
    J. $5(2)^2 - 3(2) + 6 = 20$
    K. $6(2)^2 + 3(2) - 13 = 17$

Since F, H, J, and K each came out incorrect, the answer is choice **G**.

Important note: G is **not** the correct answer simply because it is equal to 7. It is correct because all four of the other choices are **not** 7. **You absolutely must check all five choices!**

45

* **Algebraic solution:** We need to subtract $(5x^2 - 8) - (x^2 + 3x - 5)$. We first eliminate the parentheses by distributing the minus sign:

$$5x^2 - 8 - x^2 - 3x + 5$$

Finally, we combine like terms to get $4x^2 - 3x - 3$, choice **G**.

Remark: Pay careful attention to the minus and plus signs in the solution above. In particular, make sure you are distributing correctly.

45. For which nonnegative value of a is the expression $\dfrac{1}{16-a^2}$ undefined?

    A. 0
    B. 4
    C. 16
    D. 32
    E. 64

* **Solution by plugging in answer choices:** We want to find a nonnegative value for a that makes the denominator of the fraction zero. Normally we would start with choice C, but in this case it's pretty easy to see that choice B will work. Indeed, $16 - 4^2 = 16 - 16 = 0$. So the answer is choice **B**.

Algebraic solution: The expression is undefined when the denominator is zero. So we need to solve the equation $16 - a^2 = 0$. Factoring the left hand side gives the equation $(4 - a)(4 + a) = 0$. So $4 - a = 0$ or $4 + a = 0$. Therefore, we have $a = 4$ or $a = -4$. Since the question is asking for the nonnegative value of a, we choose $a = 4$, choice **B**.

Notes: (1) The given expression is a **rational function**. A rational function is a quotient of polynomials (one polynomial divided by another polynomial). A rational function is undefined when the denominator is zero.

(2) The expression $16 - a^2$ is the **difference of two squares**. In general, the difference of two squares $x^2 - y^2$ factors as $(x - y)(x + y)$.

(3) We can also solve the equation $16 - a^2 = 0$ by adding a^2 to each side of the equation to get $16 = a^2$, and then using the **square root property** to get $\pm 4 = a$.

Note that the **square root property** says that if $x^2 = k^2$, then $x = \pm k$. This is different from taking a square root since it leads to two solutions.

46. What is the greatest integer x that satisfies the inequality $2 + \frac{x}{5} < 7$?

    F. 20
    G. 22
    H. 24
    J. 25
    K. 26

Solution by plugging in answer choices: Since the word "greatest" appears in the problem, let's start with the largest answer choice, choice K. Now $2 + \frac{26}{5} = 7.2$ (use your calculator). This is too big, so we can eliminate choice K.

Let's try choice J next. We have $2 + \frac{25}{5} = 7$. This is just barely too big, so the answer is choice **H.**

* Algebraic solution: Let's solve the inequality. We start by subtracting 2 from each side of the given inequality to get $\frac{x}{5} < 5$. We then multiply each side of this inequality by 5 to get $x < 25$. The greatest integer less than 25 is 24, choice **H.**

47. Which of the following expressions is equivalent to $\frac{5k+50}{5}$?

    A. $k + 10$
    B. $k + 50$
    C. $7k + 10$
    D. $11k$
    E. $50k$

Solution by plugging in answer choices: Let's choose a value for k, say $k = 2$. We first substitute a 2 in for k into the given expression and use our calculator. We type in the following: (5*2 + 50) / 5 to get $k = 12$. Put a nice big, dark circle around this number so that you can find it easily later. We now substitute a 2 into each answer choice.

    A. 12
    B. 52
    C. 24
    D. 22
    E. 100

We now compare each of these numbers to the number that we put a nice big, dark circle around. Since B, C, D and E are incorrect we can eliminate them. Therefore, the answer is choice **A.**

Important note: A is **not** the correct answer simply because it is equal to 12. It is correct because all four of the other choices are **not** 12. **You absolutely must check all five choices!**

Algebraic solution: Most students have no trouble at all adding two fractions with the same denominator. For example,

$$\frac{5k}{5} + \frac{50}{5} = \frac{5k + 50}{5}$$

But these same students have trouble reversing this process.

$$\frac{5k + 50}{5} = \frac{5k}{5} + \frac{50}{5}$$

Note that these two equations are **identical** except that the left and right hand sides have been switched. Note also that to break a fraction into two (or more) pieces, the original denominator is repeated for **each** piece.

* An algebraic solution to the above problem consists of the following quick computation

$$\frac{5k + 50}{5} = \frac{5k}{5} + \frac{50}{5} = k + 10$$

This is choice **A.**

48. What values for x are solutions for $x^2 - 2x = 15$?

    F. 3 and 5
    G. 0 and 5
    H. −3 and 0
    J. −3 and −5
    K. −3 and 5

Solution by plugging in answer choices: Let's start with choice H and try $x = 0$ (since it's easier to plug in than −3). We have $0^2 - 2(0) = 0$. Since this is not 15 we can eliminate choices H and G.

Let's go to choice F next and try $x = 5$. We have $5^2 - 2(5) = 25 - 10 = 15$. Since this works the answer is either choice F or choice K.

Let's try $x = 3$. We have $3^2 - 2(3) = 9 - 6 = 3$. Since this is not 15 we can eliminate choice A. The answer is therefore choice **K.**

Remark: We should try $x = -3$ to be safe: $(-3)^2 - 2(-3) = 9 + 6 = 15$. This works, so the answer is in fact choice **K**.

* **Algebraic solution:** We subtract 15 from each side of the given equation to get $x^2 - 2x - 15 = 0$. We can factor the left hand side to get $(x - 5)(x + 3) = 0$. So $x - 5 = 0$ or $x + 3 = 0$. Therefore, $x = 5$ or $x = -3$, choice **K**.

LEVEL 2: GEOMETRY

49. A rectangle has a perimeter of 16 meters and an area of 15 square meters. What is the longest of the side lengths, in meters, of the rectangle?

    A. 3
    B. 5
    C. 10
    D. 15
    E. 16

* **Solution by plugging in answer choices:** Let's start with choice C and guess that the longest side of the rectangle is 10 meters long. But then the length of the two longer sides of the rectangle adds up to 20 meters which is greater than the perimeter. So we can eliminate choices C, D, and E.

Let's try choice B next. So we are guessing that the longest side of the rectangle is 5. Since the perimeter is 16, it follows that the shortest side must have length 3 (see Remark (1) below for more clarification). So the area is $(5)(3) = 15$. Since this is correct, the answer is choice **B**.

Remarks: (1) If one side of the rectangle has a length of 5 meters, then the opposite side also has a length of 5 meters. Since the perimeter is 16 meters, this leaves $16 - 5 - 5 = 6$ meters for the other two sides. It follows that a shorter side of the rectangle has length $\frac{6}{2} = 3$ meters.

(2) When guessing the longest side of the rectangle we can use the area instead of the perimeter to find the shortest side. For example, if we guess that the longest side is 5, then since the area is 15 it follows that the shortest side is 3. We would then check to see if we get the right perimeter. In this case we have $P = 2(5) + 2(3) = 16$ which is correct.

Algebraic solution: We are given that $2x + 2y = 16$ and $xy = 15$. If we divide each side of the first equation by 2, we get $x + y = 8$. Subtracting each side of this equation by x, we get $y = 8 - x$.

We now replace y by $8 - x$ in the second equation to get $x(8 - x) = 15$. Distributing the x on the left yields $8x - x^2 = 15$. Subtracting $8x$ and adding x^2 to each side of this equation gives us $0 = x^2 - 8x + 15$. The right hand side can be factored to give $0 = (x - 5)(x - 3)$. So we have $x - 5 = 0$ or $x - 3 = 0$. So $x = 5$ or $x = 3$. Since the question asks for the longest of the side lengths, the answer is $x = 5$, choice **B.**

Here is a picture for extra clarification.

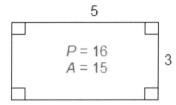

50.   In the figure below, $\angle MAT$ measures 100°, $\angle AMT$ measures 20°, and points A, T, and H are collinear. What is the measure of $\angle MTH$?

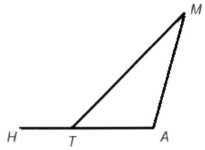

F.   100°
G.   110°
H.   120°
J.   130°
K.   140°

* Let's add the given information to the picture.

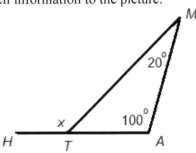

50

There are now two ways to proceed:

Method 1: The measure of an exterior angle of a triangle is the sum of the measures of the two opposite interior angles of the triangle. So we have $x = 20 + 100 = 120$, choice **H.**

Method 2: Since the angle measures of a triangle sum to 180, angle MTA measures $180 - 20 - 100 = 60°$. Since angles *MTH* and *MTA* form a **linear pair**, they are **supplementary**. Therefore, $x = 180 - 60 = 120$, choice **H.**

Note: Supplementary angles have measures which add to 180°.

51. In the standard (x, y) coordinate plane, what is the slope of the line segment joining the points (3,–5) and (7,2) ?

    A. $-\dfrac{7}{4}$

    B. $-\dfrac{3}{4}$

    C. $\dfrac{4}{7}$

    D. $\dfrac{3}{4}$

    E. $\dfrac{7}{4}$

Solution by drawing a picture: Let's plot the two points.

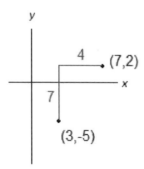

Note that to get from (3,–5) to (7,2) we move up 7 and right 4. Therefore, the answer is $\dfrac{7}{4}$, choice **E.**

Note: If you cannot see where the 7 and 4 come from visually, then you can formally find the differences: $2 - (-5) = 7$ and $7 - 3 = 4$.

51

* Solution using the slope formula: $\frac{2-(-5)}{7-3} = \frac{7}{4}$, choice **E.**

Notes:

$$\text{Slope} = m = \frac{\text{rise}}{\text{run}} = \frac{y_2 - y_1}{x_2 - x_1}$$

Lines with positive slope have graphs that go upwards from left to right. Lines with negative slope have graphs that go downwards from left to right. If the slope of a line is zero, it is horizontal. Vertical lines have **no** slope (this is different from zero slope).

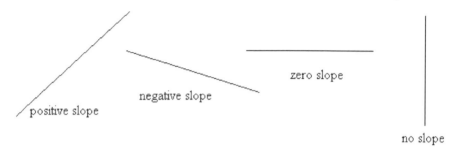

52. What is the slope of any line perpendicular to the line $3x - 2y = 5$?

F. -2

G. $-\frac{3}{5}$

H. $-\frac{2}{3}$

J. 3

K. 5

* Let's first find the slope of the given line. We solve for y:

$$3x - 2y = 5$$
$$-2y = -3x + 5$$
$$y = \frac{3}{2}x - \frac{5}{2}$$

So we see that the slope of the given line is $\frac{3}{2}$. Therefore the slope of any line perpendicular to the given line is $-\frac{2}{3}$, choice **H.**

Clarification of the algebra: To get from the first equation to the second equation we subtracted $3x$ from each side.

52

To get from the second equation to the third equation we divided each side of the equation by –2. Note that we had to divide each term on the right hand side by –2. So we have $\frac{-3x}{-2} = \frac{3x}{2} = \frac{3}{2}x$ and $\frac{5}{-2} = -\frac{5}{2}$.

Remarks: (1) Perpendicular lines have slopes that are negative reciprocals of each other. The reciprocal of $\frac{3}{2}$ is $\frac{2}{3}$. The negative reciprocal of $\frac{3}{2}$ is $-\frac{2}{3}$.

(2) The equation of a line in **slope-intercept** form is $y = mx + b$ where m is the slope of the line and $(0, b)$ is the y-intercept of the line.

In the above solution we put the equation into slope-intercept form by solving for y.

53. The coordinates of the endpoints of $\overline{AB}$, in the standard (x, y) coordinate plane, are $(-3,-5)$ and $(3,13)$. What is the y-coordinate of the midpoint of $\overline{AB}$?

    A. 0
    B. 2
    C. 4
    D. 9
    E. 10

* We simply take the average of the y-coordinates of the two given points to get $\frac{-5+13}{2} = \frac{8}{2} = 4$, choice **C**.

Here is a picture of the line segment.

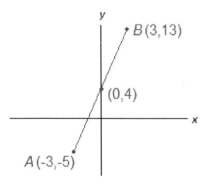

54. What is the length, in meters, of the hypotenuse of a right triangle with legs that are 5 meters long and 8 meters long, respectively?

F. $\sqrt{89}$
G. $\sqrt{91}$
H. 13
J. 20
K. 40

* Solution by Pythagorean Theorem: By the Pythagorean Theorem, we have $c^2 = 5^2 + 8^2 = 25 + 64 = 89$ where c is the length of the hypotenuse of the triangle. So $c = \sqrt{89}$, choice **F.**

Remarks: (1) The Pythagorean Theorem says that if a right triangle has legs of length a and b, and a hypotenuse of length c, then $c^2 = a^2 + b^2$.

(3) The equation $c^2 = 89$ would normally have two solutions: $c = \sqrt{89}$, and $c = -\sqrt{89}$. But the length of a side of a triangle cannot be negative, so we reject $-\sqrt{89}$.

55. Rectangle $PQRS$ has vertices $P(1,3)$, $Q(1,-2)$, and $S(5,3)$. What is the y-coordinate of vertex R ?

A. -3
B. -2
C. -1
D. 0
E. 1

* Solution by drawing a picture: Let's a draw a picture of the rectangle

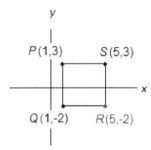

We see that the y-coordinate of point R is –2, choice **B.**

54

56. Four points, $X, Y, Z,$ and W, lie on a circle having a circumference of 30 units. Y is 4 units counterclockwise from X. Z is 10 units clockwise from X. W is 14 units clockwise from X and 16 units counterclockwise from X. What is the order of the points, starting with X and going clockwise around the circle?

    F. X, Y, Z, W
    G. X, Y, W, Z
    H. X, Z, Y, W
    J. X, Z, W, Y
    K. X, Z, Y, W

* Solution by drawing a picture: Let's draw a picture of the circle.

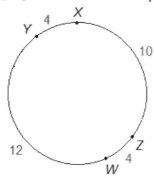

From the picture we see that the order of the points, starting with X, and going clockwise around the circle is X, Z, W, Y. This is choice **J**.

Remark: It was not necessary to write down the arc lengths in the picture. I just did it for completeness.

LEVEL 2: PROBABILITY AND STATISTICS

57. The Math Club needs to select 4 officers by first selecting the president, then the vice president, then the secretary, and finally the treasurer. If there are 20 members who are eligible to hold office and no member can hold more than 1 office, which of the following gives the number of different possible results of the election?

    A. $20 \cdot 19 \cdot 18 \cdot 17$
    B. $19 \cdot 18 \cdot 17 \cdot 16$
    C. 20^4
    D. 19^4
    E. 15^4

Solution using the counting principle: There are 20 possible choices for the president. After selecting the president, there are 19 possible choices for the vice president. After choosing the vice president, there are 18 possible choices for the secretary. Finally, after choosing the secretary, there 17 possible choices for the treasurer. By the counting principle we get $20 \cdot 19 \cdot 18 \cdot 17$ possible results, choice **A**.

See the remark at the end of problem 26 for more information on the counting principle.

* Solution using permutations: There are 20 members, and we are selecting 4 of them in a specific order. It follows that there are $_{20}P_4 = 20 \cdot 19 \cdot 18 \cdot 17$ possible results, choice **A**.

Permutations: $_{20}P_4$ means the number of **permutations** of 20 things taken 4 at a time. In a permutation order matters (as opposed to the **combination** $_{20}C_4$ where the order does not matter).

$$_{20}P_4 = \frac{20!}{16!} = 20 \cdot 19 \cdot 18 \cdot 17.$$

In general, if n is an integer, then $n! = 1 \cdot 2 \cdot 3 \cdots n$

If n and k are integers, then $_nP_r = \frac{n!}{(n-r)!}$

$0! = 1$ by definition.

Note that on the ACT you **do not** need to know the permutation formula. If necessary, you can do this computation very quickly on your graphing calculator. To compute $_{20}P_4$, type 20 into your calculator, then in the **Math** menu scroll over to **Prb** and select **nPr** (or press **2**). Then type 4 and press **Enter**. You will get an answer of 116,280.

Note that in this problem we did not actually need to perform the computation.

58. What is the median of the data given below?

$$45, 62, 71, 36, 22, 48, 51, 26, 62, 39$$

F. 35
G. 46
H. 46.2
J. 46.5
K. 62

56

* Let's rewrite the list of numbers in increasing order.

$$22, 26, 36, 39, 45, 48, 51, 62, 62, 71$$

There are 10 numbers in the list. Since 10 is even, we take the average of the 5th and 6th numbers. We get $\frac{45+48}{2} = \frac{93}{2} = 46.5$, choice **J**.

59. Kenneth's test average after 7 tests was 81. His score on the 8th test was 92. If all 8 tests were equally weighted, which of the following is closest to his test average after 8 tests?

    A. 80
    B. 82
    C. 84
    D. 86
    E. 88

* Solution by changing the average to a sum: We change the average to a sum using the formula

$$\text{Sum} = \text{Average} \cdot \text{Number}$$

We are averaging seven numbers. Thus, the **Number** is 7. The **Average** is given to be 81. So the **Sum** of the seven numbers is $81 \cdot 7 = 567$.

When we add 92, the **Sum** becomes $567 + 92 = 659$. The **Average** is then $\frac{659}{8} = 82.375$. This is closest to 82, choice **B**.

60. Of the marbles in a jar, 14 are orange. Sarah randomly takes one marble out of the jar. If the probability is $\frac{2}{7}$ that the marble she chooses is orange, how many marbles are in the jar?

    F. 4
    G. 14
    H. 28
    J. 49
    K. 98

Solution by plugging in answer choices: Let's start with choice H and guess that there are 28 marbles in the jar. We have $\frac{2}{7} \cdot 28 = 8$. This is too small. So we can eliminate choices F, G, and H.

Let's try choice J next and guess that there are 49 marbles in the jar. We have $\frac{2}{7} \cdot 49 = 14$. This is correct. So the answer is choice **J**.

Remarks: (1) Saying that the probability is $\frac{2}{7}$ that an orange marble will be chosen is equivalent to saying that $\frac{2}{7}$ of the marbles are orange. So for example, when we guess that there are 49 marbles in the jar, we need to compute $\frac{2}{7}$ of 49. The word "of" always means multiplication.

(2) All of the computations above can be done by hand or in your calculator. For example, to compute $\frac{2}{7} \cdot 28$, simply type 2 / 7 * 28 followed by ENTER. The output will be **8**.

* **Algebraic solution:** Let x be the total number of marbles in the jar. We are given that $\frac{2}{7}x = 14$. We multiply each side of this equation by $\frac{7}{2}$ to get that $x = 14 \cdot \frac{7}{2} = 49$, choice **J**.

LEVEL 2: TRIGONOMETRY

61. For $\angle R$ in $\triangle PQR$ below, which of the following trigonometric expressions has value $\frac{5}{12}$?

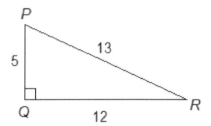

A. tan R
B. cot R
C. sin R
D. csc R
E. sec R

Solution by plugging in answer choices: Let's start with choice C and compute sin $R = \frac{\text{OPP}}{\text{HYP}} = \frac{5}{13}$. This is incorrect, but the numerator is correct. How do we make the denominator 12 instead of 13? Well we use tangent instead. Indeed, tan $R = \frac{\text{OPP}}{\text{ADJ}} = \frac{5}{12}$. So the answer is choice **A**.

Note: In the above solution, OPP stands for "opposite," ADJ stands for "adjacent," and HYP stands for "hypotenuse."

* **Quick solution:** Note that the numerator and denominator of the fraction are the lengths of the legs of the right triangle. So the answer is most likely a tangent or cotangent. Choices A and D look like the only candidates. Now we simply check: $\tan R = \dfrac{\text{OPP}}{\text{ADJ}} = \dfrac{5}{12}$. So the answer is choice **A.**

Here is a quick lesson in **right triangle trigonometry** for those of you that have forgotten.

Let's begin by focusing on angle A in the following picture:

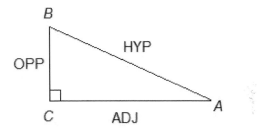

Note that the **hypotenuse** is ALWAYS the side opposite the right angle.

The other two sides of the right triangle, called the **legs**, depend on which angle is chosen. In this picture we chose to focus on angle A. Therefore, the opposite side is BC, and the adjacent side is AC.

Now you should simply memorize how to compute the six trig functions:

$$\sin A = \frac{\text{OPP}}{\text{HYP}} \qquad \csc A = \frac{\text{HYP}}{\text{OPP}}$$

$$\cos A = \frac{\text{ADJ}}{\text{HYP}} \qquad \sec A = \frac{\text{HYP}}{\text{ADJ}}$$

$$\tan A = \frac{\text{OPP}}{\text{ADJ}} \qquad \cot A = \frac{\text{ADJ}}{\text{OPP}}$$

Here are a couple of tips to help you remember these:

(1) Many students find it helpful to use the word SOHCAHTOA. You can think of the letters here as representing sin, opp, hyp, cos, adj, hyp, tan, opp, adj.

(2) The three trig functions on the right are the reciprocals of the three trig functions on the left. In other words, you get them by interchanging the numerator and denominator. It's pretty easy to remember that the reciprocal of tangent is cotangent. For the other two, just remember that the "s" goes with the "c" and the "c" goes with the "s." In other words, the reciprocal of sine is cosecant, and the reciprocal of cosine is secant.

To make sure you understand this, compute all six trig functions for each of the angles (except the right angle) in the triangle given in problem 61. Please try this yourself before looking at the answers below.

$$\sin P = \frac{12}{13} \qquad \csc P = \frac{13}{12} \qquad \sin R = \frac{5}{13} \qquad \csc R = \frac{13}{5}$$

$$\cos P = \frac{5}{13} \qquad \sec P = \frac{13}{5} \qquad \cos R = \frac{12}{13} \qquad \sec R = \frac{13}{12}$$

$$\tan P = \frac{12}{5} \qquad \cot P = \frac{5}{12} \qquad \tan R = \frac{5}{12} \qquad \cot R = \frac{12}{5}$$

62. The figure below shows a right triangle whose hypotenuse is 4 feet long. How many feet long is the shorter leg of this triangle?

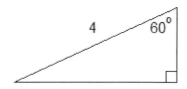

F. 2

G. 8

H. $2\sqrt{3}$

J. $\frac{2\sqrt{3}}{3}$

K. $\frac{8\sqrt{3}}{3}$

* The shorter leg of the triangle is adjacent to the 60° angle. So we will use cosine. We have $\cos 60° = \frac{ADJ}{HYP} = \frac{ADJ}{4}$. So ADJ = 4cos 60° = $4(\frac{1}{2})$ = 2, choice **F**.

Remarks: (1) If you do not see why we have $\cos 60° = \frac{ADJ}{HYP}$, review the basic trigonometry given after the solution to problem 61.

(2) To get from $\cos 60° = \frac{ADJ}{4}$ to ADJ = 4cos 60°, we simply multiply each side of the first equation by 4.

For those of you that like to cross multiply, the original equation can first be rewritten as $\frac{\cos 60°}{1} = \frac{ADJ}{4}$.

(3) There are several ways to compute cos 60°. The easiest is to simply put it into your calculator. The output will be .5.

(4) Make sure that your calculator is in degree mode before using your calculator. Otherwise you will get an incorrect answer.

If you are using a TI-84 (or equivalent) calculator press MODE and on the third line, make sure that DEGREE is highlighted. If it is not, scroll down and select it. If possible, do not alter this setting until you are finished taking your ACT.

(5) For the ACT, it is worth knowing the following two special triangles:

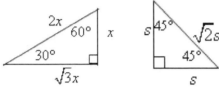

Some students get a bit confused because there are variables in these pictures. But the pictures become simplified if we substitute a 1 in for the variables. Then the sides of the 30, 60, 90 triangle are 1, 2 and $\sqrt{3}$ and the sides of the 45, 45, 90 triangle are 1, 1 and $\sqrt{2}$. The variable just tells us that if we multiply one of these sides by a number, then we have to multiply the other two sides by the same number. For example, instead of 1, 1 and $\sqrt{2}$, we can have 3, 3 and $3\sqrt{2}$ (here $s = 3$), or $\sqrt{2}, \sqrt{2}$, and 2 (here $s = \sqrt{2}$). For this problem, we have $\cos 60° = \dfrac{\text{ADJ}}{\text{HYP}} = \dfrac{x}{2x} = \dfrac{1}{2}$.

63. In the right triangle pictured below, $a, b,$ and c are the lengths of its sides. What is the value of $\cos A$?

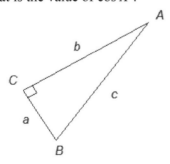

A. $\dfrac{c}{b}$

B. $\dfrac{a}{b}$

C. $\dfrac{a}{c}$

D. $\dfrac{b}{a}$

E. $\dfrac{b}{c}$

* $\cos A = \dfrac{\text{ADJ}}{\text{HYP}} = \dfrac{b}{c}$, choice **E**.

Remark: If you do not see why we have $\cos A = \dfrac{\text{ADJ}}{\text{HYP}}$, review the basic trigonometry given after the solution to problem 61.

64. As shown below, a 10-foot ramp forms an angle of 23° with the ground, which is horizontal. Which of the following is an expression for the vertical rise, in feet, of the ramp?

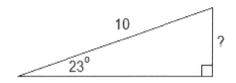

F. 10 cos 23°

G. 10 sin 23°

H. 10 tan 23°

J. 10 cot 23°

K. 10 sec 23°

* We have $\sin 23° = \dfrac{\text{OPP}}{\text{HYP}} = \dfrac{\text{OPP}}{10}$. So OPP = 10sin 23°, choice **G**.

Remarks: (1) If you do not see why we have $\sin 23° = \dfrac{\text{OPP}}{\text{HYP}}$, review the basic trigonometry given after the solution to problem 61.

(2) To get from $\sin 23° = \dfrac{\text{OPP}}{10}$ to OPP = 10sin 23°, we simply multiply each side of the first equation by 10.

For those of you that like to cross multiply, the original equation can first be rewritten as $\dfrac{\sin 23°}{1} = \dfrac{\text{OPP}}{10}$.

LEVEL 3: NUMBER THEORY

65. In what order should $\frac{7}{3}$, $\frac{9}{4}$, $\frac{11}{5}$, and $\frac{17}{8}$ be listed to be arranged in increasing order?

A. $\frac{7}{3} < \frac{9}{4} < \frac{11}{5} < \frac{17}{8}$

B. $\frac{9}{4} < \frac{11}{5} < \frac{17}{8} < \frac{7}{3}$

C. $\frac{11}{5} < \frac{17}{8} < \frac{9}{4} < \frac{7}{3}$

D. $\frac{17}{8} < \frac{11}{5} < \frac{7}{3} < \frac{9}{4}$

E. $\frac{17}{8} < \frac{11}{5} < \frac{9}{4} < \frac{7}{3}$

* Solution by changing fractions to decimals: We divide in our calculator to get

$$\frac{7}{3} \approx 2.33 \qquad \frac{9}{4} = 2.25 \qquad \frac{11}{5} = 2.2 \qquad \frac{17}{8} = 2.125$$

We have $2.125 < 2.2 < 2.25 < 2.33$. Therefore $\frac{17}{8} < \frac{11}{5} < \frac{9}{4} < \frac{7}{3}$, choice **E.**

Notes: (1) Decimals are much easier to compare than fractions, and changing fractions to decimals is easy. Just divide the numerator by the denominator in your calculator.

(2) We can compare two decimals by looking at the first position where they disagree. For example, 2.125 is less than 2.2 because 1 is less than 2. If a digit is missing, there is a hidden 0 there. Thus 2.2 is less than 2.25 because 2.2 is the same as 2.20 and 0 is less than 5.

(3) Comparing the original fractions would be quite time consuming. As an example, let's compare $\frac{7}{3}$ and $\frac{9}{4}$. One of the quicker ways to do this would be to perform a cross multiplication. We start by putting "?" between the two numbers. This "?" represents either "<" or ">," but we do not know which yet. We then multiply each side of this (unknown) inequality by 12 as follows

$$\frac{7}{3} \, ? \, \frac{9}{4} \qquad 12(\tfrac{7}{3}) \, ? \, (\tfrac{9}{4})12 \qquad 4 \cdot 7 \, ? \, 9 \cdot 3 \qquad 28 \, ? \, 27$$

So we see that the correct symbol is >, that is $\frac{7}{3} > \frac{9}{4}$.

63

66. Jessica is planning to bake a pie using a recipe that requires $5\frac{1}{2}$ tablespoons of cinnamon. She only has $2\frac{3}{4}$ tablespoons left. The amount she has left is what fraction of the amount of cinnamon she needs for the recipe?

F. $\frac{3}{4}$

G. $\frac{1}{2}$

H. $\frac{1}{4}$

J. $\frac{1}{5}$

K. $\frac{1}{8}$

Quick calculator computation: We type the following into our TI-84 calculator: $(2 + 3 / 4) / (5 + 1 / 2)$ ENTER MATH ENTER ENTER.

The output is $\frac{1}{2}$, choice **G.**

Alternative: We can also do this computation with decimals: 2.75 / 5.5. The output is .5 which is the same as $\frac{1}{2}$, choice **G.**

* Mental Computation: Observe that when we double $2\frac{3}{4}$, we get $5\frac{1}{2}$. So the answer is choice **G.**

Hand solution: Let's change the mixed numerals to improper fractions. We have $2\frac{3}{4} = \frac{11}{4}$ and $5\frac{1}{2} = \frac{11}{2}$. So $\frac{11}{4} \div \frac{11}{2} = \frac{11}{4} \cdot \frac{2}{11} = \frac{1}{2}$, choice **G.**

Notes: (1) A quick way to change a mixed numeral to an improper fraction is to multiply the denominator with the whole number and then add the numerator. This becomes the numerator of the improper fraction. For example, for $2\frac{3}{4}$, we have $4 \cdot 2 + 3 = 8 + 3 = 11$. So the improper fraction is $\frac{11}{4}$.

67. Which of the following is a rational number?

A. $\sqrt{3}$

B. $\sqrt{\pi^2}$

C. $\sqrt{11}$

D. $\sqrt{\frac{9}{64}}$

E. $\sqrt{\frac{49}{5}}$

64

Calculator method: Your TI-84 calculator can turn most rational numbers into fractions. So simply input each number into your calculator and then press MATH ENTER ENTER (beginning with choice C). When you get to choice D, your calculator will give $\frac{3}{8}$. The answer is choice **D**.

Remarks: (1) A rational number is a number that can be expressed as a fraction. More specifically, a rational number has the form $\frac{a}{b}$, where a and b are integers with $b \neq 0$.

(2) The decimal expansion of a rational number either terminates or repeats. In this example, when we put $\sqrt{\frac{9}{64}}$ into our calculator we get 0.375. Since this is a terminating decimal, $\sqrt{\frac{9}{64}}$ is a rational number.

* Quick observation: Simply observe that both 9 and 64 are perfect squares. Therefore $\sqrt{\frac{9}{64}}$ is a rational number.

Notes: (1) A **perfect square** is an integer that is equal to the square of another integer. For example, 9 is a perfect square because $9 = 3^2$.

(2) $\sqrt{\frac{9}{64}} = \frac{\sqrt{9}}{\sqrt{64}} = \frac{3}{2}$. Here we have used the following property of square roots:

$$\sqrt{\frac{a}{b}} = \frac{\sqrt{a}}{\sqrt{b}}$$

68. A recipe for a sports drink calls for 7 parts fruit juice to 3 parts water. To make 40 liters of this drink, how many liters of fruit juice should be used?

  A. 4
  B. 10
  C. 12
  J. 28
  K. 40

* We can represent the number of liters of fruit juice by $7x$ and the number of liters of water by $3x$ for some number x. Then the total amount of liquid is $10x$ which must be equal to 40. We have that $10x = 40$ implies $x = 4$. Since we want the number of liters of fruit juice, we need to find $7x$. This is $7(4) = 28$, choice **J**.

65

Important note: After you find x make sure you look at what the question is asking for. A common error is to give an answer of 4. But the amount of fruit juice is **not** equal to x. It is equal to $7x$.

Alternate solution: We set up a ratio of the amount of fruit juice to the total liquid, then we cross multiply and divide.

$$\begin{array}{ccc} \text{fruit juice} & 7 & x \\ \text{total liquid} & 10 & 40 \end{array}$$

$$\frac{7}{10} = \frac{x}{40}$$

$$10x = 280$$

$$x = \frac{280}{10} = 28.$$

This is choice **J**.

69. An integer is decreased by 30% and the resulting number is then increased by 35%. The final number is what percent of the original number?

  A. 90
  B. 92
  C. 94.5
  D. 100
  E. 105

* **Solution by picking a number:** Since there are percents in the problem we pick the number 100. When we decrease 100 by 30% we get 70. To increase 70 by 35 we multiply $1.35(70) = 94.5$, choice **C**.

Notes: (1) Decreasing 100 by 30% is the same as decreasing 100 by 30. This works **only** with the number 100.

(2) To increase 70 by 35% it would be **wrong** to add 35 to 70 because 35% of 70 is **not** 35. To take 35% of 70 we would multiply 70 by 0.35 to get $(0.35)(70) = 24.5$, Thus, increasing 70 by 35% gives $70 + 24.5 = 94.5$. In the solution above we did this even faster by multiplying 70 by 1.35. This is the same because $70 + (0.35)(70) = (1 + 0.35)(70) = 1.35(70)$ by the **distributive property**.

(3) See problem 3 if you need to review changing percents to decimals.

Direct solution: Let x be the original number. We decrease x by 30% to get $0.7x$. We then increase $0.7x$ by 35% to get $(1.35)(0.7)x = 0.945x$. So the final number is 94.5% of the original number, choice **C**.

Notes: (1) Decreasing a number by 30% is the same as taking 70% of that number. So we can decrease a number by 30% by multiplying it by 0.7.

(2) We can also decrease x by 30% by subtracting $0.3x$ from x to get $x - 0.3x = (1 - 0.3)x = 0.7x$.

(3) To increase x by 35% we can multiply x by 1.35 or we can also add $0.35x$ to x to get $x + 0.35x = (1 + 0.35)x = 1.35x$.

(4) To change the decimal 0.945 to a percent we simply move the decimal point to the right 2 places.

70. Karen, Lisa, and Maria shared a pizza pie. Karen ate $\frac{1}{4}$ of the pie, Lisa ate $\frac{1}{3}$ of the pie, and Maria ate the rest. What is the ratio of Karen's share to Lisa's share to Maria's share?

    F. 3:4:5
    G. 3:4:6
    H. 4:3:5
    J. 4:3:6
    K. 5:4:3

* Solution by changing the fractional parts to wholes: Let's split the pie into $(3)(4) = 12$ pieces. Then Karen ate $\frac{1}{4}(12) = 3$ pieces and Lisa ate $\frac{1}{3}(12) = 4$ pieces. Therefore, Maria ate $12 - 3 - 4 = 5$ pieces. So the ratio is 3:4:5, choice **F**.

Note: We chose to split the pie into 12 pieces because 12 is the product of the two denominators.

The product is usually a good choice. The least common denominator may sometimes give a quicker solution, but when in doubt just go with the product (in this problem, the product of the two denominators is the same as the least common denominator).

71. John drove 8 hours from New York to Virginia. The total distance he travelled was 535 miles, and he averaged 65 miles per hour for the first 3 hours. Which of the following is closest to his average driving speed, in miles per hour, for the remainder of his drive?

    A. 60
    B. 62
    C. 65
    D. 66
    E. 68

* We use the formula **distance = rate · time** (or $d = rt$ for short). During the first 3 hours John travelled $d = rt = (65)(3) = 195$ miles. This leaves $535 - 195 = 340$ miles for the remaining $8 - 3 = 5$ hours. So his average driving speed for the remainder of his drive is $r = \frac{d}{t} = \frac{340}{5} = 68$ m/h, choice **E**.

72. Which of the following is the least common denominator for the expression below?
$$\frac{1}{11^2 \cdot 29} + \frac{1}{11 \cdot 19^2 \cdot 29} + \frac{1}{11 \cdot 29^4}$$

    F. $11 \cdot 19$
    G. $11 \cdot 29$
    H. $11 \cdot 19 \cdot 29$
    J. $11^2 \cdot 19^2 \cdot 29^4$
    K. $11^4 \cdot 19^2 \cdot 29^6$

* We are looking for the **least common multiple of** $11^2 \cdot 29$, $11 \cdot 19^2 \cdot 29$, and $11 \cdot 29^4$. We take the highest power of each prime that appears to get $11^2 \cdot 19^2 \cdot 29^4$, choice **J**.

Remark: See problem 34 for more detailed information on computing least common multiples.

LEVEL 3: ALGEBRA AND FUNCTIONS

73. For all x, $(5x - 4)^2 = ?$

    A. $10x - 8$
    B. $10x^2 - 8$
    C. $25x^2 + 16$
    D. $25x^2 - 20x + 16$
    E. $25x^2 - 40x + 16$

Solution by picking a number: Let's choose a value for x, say $x = 2$. Then

$$(5x - 4)^2 = (5 \cdot 2 - 4)^2 = (10 - 4)^2 = 6^2 = \mathbf{36}.$$

Put a nice big dark circle around **36** so you can find it easier later. We now substitute 2 for x into each answer choice:

A. $10(2) - 8 = 12$
B. $10(2)^2 - 8 = 32$
C. $25(2)^2 + 16 = 116$
D. $25(2)^2 - 20(2) + 16 = 76$
E. $25(2)^2 - 40(2) + 16 = 36$

Since A, B, C, and D each came out incorrect, the answer is choice **E**.

Important note: E is **not** the correct answer simply because it is equal to 36. It is correct because all four of the other choices are **not** 36. **You absolutely must check all five choices!**

* Algebraic solution:

$$(5x - 4)^2 = (5x - 4)\,(5x - 4) = 25x^2 - 20x - 20x + 16 = 25x^2 - 40x + 16$$

This is choice **E**.

Remarks: (1) To square an expression means to multiply that expression by itself. A common mistake is to write $(5x - 4)^2 = 25x^2 + 16$. This is not correct.

(2) There are several ways to multiply two binomials. One way familiar to many students is by FOILing. If you are comfortable with the method of FOILing you can use it here, but an even better way is to use the same algorithm that you already know for multiplication of whole numbers.

Here is how the algorithm works. Try to understand it yourself before reading the full explanation below.

$$
\begin{array}{r}
5x - 4 \\
5x - 4 \\
\hline
-20x + 16 \\
25x^2 - 20x + 0 \\
\hline
25x^2 - 40x + 16
\end{array}
$$

What we did here is mimic the procedure for ordinary multiplication. We begin by multiplying –4 by –4 to get 16. We then multiply –4 by $5x$ to get –20x. This is where the first row under the first line comes from.

Next we put 0 in as a placeholder on the next line. We then multiply $5x$ by –4 to get –20x. And then we multiply $5x$ by $5x$ to get $25x^2$. This is where the second row under the first line comes from.

Now we add the two rows to get $25x^2 - 40x + 16$.

74. The expression $x^2 - x - 12$ can be written as the product of two binomial factors with integer coefficients. One of the binomials is $(x + 3)$. Which of the following is the other binomial?

   F. $x^2 - 4$
   G. $x^2 + 4$
   H. $x - 4$
   J. $x + 4$
   K. $x + 5$

Solution by picking a number: Let's choose a value for x, say $x = 2$. Then $x^2 - x - 12 = 2^2 - 2 - 12 = 4 - 14 = -10$ and $(x + 3) = 2 + 3 = 5$. So the answer should be $-\dfrac{10}{5} = $ **–2**. Put a nice big dark circle around **–2** so you can find it easier later. We now substitute 2 for x into each answer choice:

   F. $2^2 - 4 = 4 - 4 = 0$
   G. $2^2 + 4 = 4 + 4 = 8$
   H. $2 - 4 = -2$
   J. $2 + 4 = 6$
   K. $2 + 5 = 7$

Since F, G, J, and K each came out incorrect, the answer is choice **H.**

Important note: H is **not** the correct answer simply because it is equal to –2. It is correct because all four of the other choices are **not** –2. **You absolutely must check all five choices!**

Notes: (1) By picking the number $x = 2$, we have changed the problem to "the number –10 can be written as a product of 5 and what other number?" This is why the answer (to this new problem) is –2.

(2) A **binomial** has two terms. For example, the two terms of $(x + 3)$ are x and 3. The two terms of $(x - 4)$ are x and 4.

70

* **Algebraic solution:** We are being asked to factor $x^2 - x - 12$. But we are also given that one of the factors is $(x + 3)$. Since $-\frac{12}{3} = -4$, the other factor must be $(x - 4)$, choice **H.**

Note: On the ACT, an expression of the form $x^2 + bx + c$ will usually factor as $(x + m)(x + n)$ where m and n are integers and $mn = c$.

In this problem, $c = -10$ and $m = 5$. So $n = -\frac{10}{5} = -2$.

75. What value of k will satisfy the equation $0.3(k + 2100) = k$?

    A.   630
    B.   900
    C. 1200
    D. 1500
    E. 1935

Solution by plugging in answer choices: Let's start with choice C and guess that $k = 1200$. Then $0.3(k + 2100) = 0.3(1200 + 2100) = 990$. This is incorrect, so we can eliminate choice C.

Let's try choice B next. So we are guessing that $k = 900$. We then have that $0.3(k + 2100) = 0.3(900 + 2100) = 900$. This is correct. So the answer is choice **B.**

* **Algebraic solution:** We distribute 0.3 on the left hand side to get $0.3k + 630 = k$. We then subtract $0.3k$ from each side of this equation to get $630 = 0.7k$. Finally, we divide each side of this last equation by 0.7 to get $900 = k$. So the answer is choice **B.**

76. The equation shown below is true for what value of a?
$$5(a - 3) - 3(a - 2) = 7a$$

    F. $-\frac{5}{9}$

    G. $-\frac{9}{5}$

    H. 0

    J. $\frac{9}{5}$

    K. $\frac{5}{9}$

* Algebraic solution: We distribute the two expressions on the left to get $5a - 15 - 3a + 6 = 7a$. We now combine like terms on the left to get $2a - 9 = 7a$. Subtracting $2a$ from each side of this equation gives us $-9 = 5a$. Finally we divide each side of this last equation by 5 to get $-\frac{9}{5} = a$. So the answer is choice **G**.

Remark: If you have too much trouble with the algebraic solution, you can also solve this problem by "plugging in the answer choices."

77. The expression $-5x^4(4x^7 - 3x^3)$ is equivalent to

    A. $-20x^{11} + 15x^7$
    B. $-20x^{11} - 15x^7$
    C. $-20x^{28} + 15x^{12}$
    D. $-20x^{28} - 15x^{12}$
    E. $-5x^8$

Solution by picking a number: Let's choose a value for x, say $x = 2$. Then $-5x^4(4x^7 - 3x^3) = -5 \cdot 2^4(4 \cdot 2^7 - 3 \cdot 2^3) = $ **–39,040**. Put a nice big dark circle around **39,040** so you can find it easily later. We now substitute 2 for x into each answer choice:

    A. $-20(2)^{11} + 15(2)^7 = -39,040$
    B. $-20(2)^{11} - 15(2)^7 = -42,880$
    C. $-20(2)^{28} + 15(2)^{12} = -5,368,647,680$
    D. $-20(2)^{28} - 15(2)^{12} = -5,368,770,560$
    E. $-5(2)^8 = -1,280$

Choice A is the only answer choice that came out correct. Therefore, the answer is choice **A**.

Important note: A is **not** the correct answer simply because it is equal to $-39,040$. It is correct because all four of the other choices are **not** – 39,040. **You absolutely must check all five choices!**

* Algebraic solution: We have

$$-5x^4(4x^7 - 3x^3) = -5x^4 \cdot 4x^7 + 5x^4 \cdot 3x^3 = -20x^{4+7} + 15x^{4+3} = -20x^{11} + 15x^7$$

This is choice **A**.

Notes: (1) As usual, make sure that you are using the distributive property correctly.

(2) For a review of the law of exponents used here see the end of the solution to problem 43.

78. The operation & is defined as $r \,\&\, s = \frac{s^2 - r^2}{r+s}$ where r and s are real numbers and $r \neq -s$. What is the value of $(-3) \,\&\, (-4)$?

F. 2
G. 1
H. 0
J. -1
K. -2

* $(-3) \,\&\, (-4) = \frac{(-4)^2 - (-3)^2}{(-3)+(-4)} = \frac{16-9}{-3-4} = \frac{7}{-7} = -1$, choice **J.**

79. If $y < |x|$, which of the following is the solution statement for y when $x = -5$?

A. $y < -5$ or $y > 5$
B. $-5 < y < 5$
C. $y < 5$
D. $y > 5$
E. y is any real number

* When $x = -5$, we have $|x| = |-5| = 5$. So the given inequality becomes $y < 5$, choice **C.**

80. A child has set up a rows of dominoes with $(b + c)$ dominoes in each row. Which of the following is an expression for the total number of dominoes that the child has set up?

F. $a + b + c$
G. $a \cdot b \cdot c$
H. $a + b \cdot c$
J. $a \cdot b + a \cdot c$
K. $a \cdot b + c$

Solution by picking numbers: Let's pick values for a, b, and c. For example, we can choose $a = 2$, $b = 3$, and $c = 5$. In this case, the child has set up 2 rows of dominoes with $3 + 5 = 8$ dominoes in each row. It follows that the total number of dominoes is $(2)(8) = \mathbf{16}$. Put a nice big dark circle around **16** so you can find it easily later. We now substitute $a = 2$, $b = 3$, and $c = 5$ into each answer choice.

F. $2 + 3 + 5 = 10$
G. $2 \cdot 3 \cdot 5 = 30$
H. $2 + 3 \cdot 5 = 2 + 15 = 17$
J. $2 \cdot 3 + 2 \cdot 5 = 6 + 10 = 16$
K. $2 \cdot 3 + 5 = 6 + 5 = 11$

Since F, G, H and K are incorrect we can eliminate them. Therefore, the answer is choice **J.**

* Algebraic solution: We multiply the number of rows of dominoes with the number of dominoes per row to get $a(b + c) = a \cdot b + a \cdot c$, choice **J.**

Note: As usual, make sure that you are using the distributive property correctly.

LEVEL 3: GEOMETRY

81. In the figure below, adjacent sides meet at right angles and the lengths given are in inches. What is the area of the figure, in square inches?

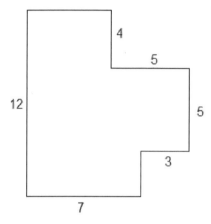

- **A.** 90
- **B.** 91
- **C.** 112
- **D.** 120
- **E.** 240

* To compute the area of the figure we break the figure up into 3 rectangles and compute the length and width of each rectangle.

74

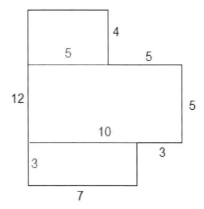

The width of the bottom rectangle is $12 - 5 - 4 = 3$ in, and the length is given to be 7 making the area $7 \cdot 3 = 21$ in^2. The length of the middle rectangle is $7 + 3 = 10$ in, and the width is given as 5 in making the area $10 \cdot 5 = 50$ in^2. The length of the top rectangle is $(7 + 3) - 5 = 5$ in, and the width is given to be 4 in making the area $5 \cdot 4 = 20$ in^2. We get the total area by adding the areas of these rectangles: $A = 21 + 50 + 20 = 91$ in^2, choice **B**.

Remark: Notice that if we have the full length of a line segment, and one partial length of the same line segment, then we get the other partial length by subtracting the two given lengths.

82. In the standard (x, y) coordinate plane, the center of the circle shown below lies on the y-axis at $y = 3$. If the circle is tangent to the x-axis, which of the following is an equation of the circle?

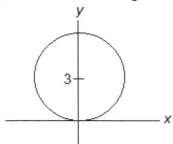

    F. $(x - 3)^2 + y^2 = 9$
    G. $x^2 + (y - 3)^2 = 9$
    H. $x^2 + (y + 3)^2 = 9$
    J. $x^2 + (y - 3)^2 = 3$
    K. $x^2 + (y + 3)^2 = 3$

* The given circle has center $(0,3)$ and a radius of 3. Therefore, the equation of the circle is $x^2 + (y-3)^2 = 9$, choice **G**.

Notes: (1) The standard form for the equation of a circle with center (h,k) and radius r is

$$(x-h)^2 + (y-k)^2 = r^2$$

(2) Any point that lies on the y-axis has an x-coordinate of 0. So in this problem, the center of the circle is $(0,3)$.

(3) Note that even though the y-coordinate of the center is positive, in the equation of the circle this translates to $(y-3)$.

(4) The **radius** of a circle is the distance from the center of the circle to **any** point on the circle. In this example we can use the point $(0,0)$. It is easy to see that the distance from $(0,3)$ to $(0,0)$ is 3. So $r=3$.

(5) Observe that in the equation of the circle we use r^2 and **not** r. This is why there is a 9 in the answer and not a 3.

83. For what value of k would the following system of equations have no solutions?

$$3x - 4y = 7$$
$$9x + 2ky = 22$$

A. -12
B. -6
C. 3
D. 6
E. 12

The system of equations

$$ax + by = c$$
$$dx + ey = f$$

has no solution if $\frac{a}{d} = \frac{b}{e} \neq \frac{c}{f}$. So we solve the equation $\frac{3}{9} = \frac{-4}{2k}$. Cross multiplying yields $6k = -36$ so that $k = \frac{-36}{6} = -6$, choice **B**.

Note: In this problem $\frac{b}{e} \neq \frac{c}{f}$. Indeed, $\frac{-4}{-6} \neq \frac{7}{22}$. This guarantees that the system of equations has no solution instead of infinitely many solutions.

* **Quick solution:** We multiply 3 by 3 to get 9. So $2k = (-4)(3) = -12$, and therefore $k = \frac{-12}{2} = -6$, choice **B**.

Here is some general theory for those of you that would like further explanation.

The **general form of an equation of a line** is $ax + by = c$ where a, b and c are real numbers. If $b \neq 0$, then the slope of this line is $m = -\frac{a}{b}$. If $b = 0$, then the line is vertical and has no slope.

Let us consider 2 such equations.

$$ax + by = c$$
$$dx + ey = f$$

(1) If there is a number r such that $ra = d$, $rb = e$, and $rc = f$, then the two equations represent the **same line**. Equivalently, the two equations represent the same line if $\frac{a}{d} = \frac{b}{e} = \frac{c}{f}$. In this case the system of equations has **infinitely many solutions**.

(2) If there is a number r such that $ra = d$, $rb = e$, but $rc \neq f$, then the two equations represent **parallel** but distinct lines. Equivalently, the two equations represent parallel but distinct lines if $\frac{a}{d} = \frac{b}{e} \neq \frac{c}{f}$. In this case the system of equations has **no solution**.

(3) Otherwise the two lines intersect in a single point. In this case $\frac{a}{d} \neq \frac{b}{e}$, and the system of equations has a **unique solution**.

These three cases are illustrated in the figure below.

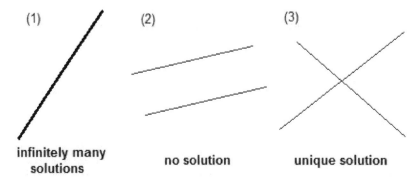

(1) infinitely many solutions

(2) no solution

(3) unique solution

Example: The following two equations represent the same line.

$$2x + 8y = 6$$
$$3x + 12y = 9$$

To see this note that $\frac{2}{3} = \frac{8}{12} = \frac{6}{9}$.(or equivalently, let $r = \frac{3}{2}$ and note that $\left(\frac{3}{2}\right)(2) = 3$, $\left(\frac{3}{2}\right)(8) = 12$, and $\left(\frac{3}{2}\right)(6) = 9$.

The following two equations represent parallel but distinct lines.

$$2x + 8y = 6$$
$$3x + 12y = 10$$

This time $\frac{2}{3} = \frac{8}{12} \neq \frac{6}{10}$.

The following two equations represent a pair of intersecting lines.

$$2x + 8y = 6$$
$$3x + 10y = 9$$

This time $\frac{2}{3} \neq \frac{8}{10}$.

84. Points P, Q, and R are vertices of an equilateral triangle. Points P, Q, and S are collinear points, with Q between P and S. What is the measure of $\angle RQS$?

F. 120°
G. 90°
H. 60°
J. 30°
K. 15°

* Let's draw a picture.

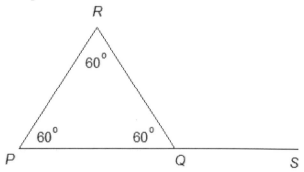

Note that in an equilateral triangle all three angles measure 60°. There are now two ways to proceed:

Method 1: The measure of an exterior angle of a triangle is the sum of the measures of the two opposite interior angles of the triangle. So we have $m\angle RQS = 60 + 60 = 120°$, choice **F.**

Method 2: Since angles *RQS* and *RQP* form a **linear pair**, they are **supplementary**. Therefore $m\angle RQS = 180 - 60 = 120°$, choice **F.**

Note: Supplementary angles have measures which add to 180°.

Here is a quick review on triangles.

A **triangle** is a two-dimensional geometric figure with three sides and three angles. The sum of the degree measures of all three angles of a triangle is 180.

A triangle is **acute** if all three of its angles measure less than 90 degrees. A triangle is **obtuse** if one angle has a measure greater than 90 degrees. A triangle is **right** if it has one angle that measures exactly 90 degrees.

Example 1:

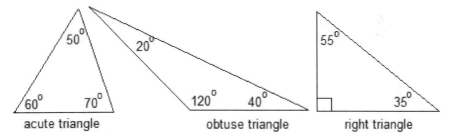

acute triangle   obtuse triangle   right triangle

A triangle is **isosceles** if it has two sides of equal length. Equivalently, an isosceles triangle has two angles of equal measure.

A triangle is **equilateral** if all three of its sides have equal length. Equivalently, an equilateral triangle has three angles of equal measure (all three angles measure 60 degrees).

Example 2:

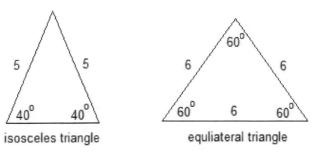

isosceles triangle   equliateral triangle

79

85. In the standard (x, y) coordinate plane, which of the following lines goes through $(0,5)$ and is perpendicular to $y = -\frac{1}{4}x + 3$?

   A. $y = -\frac{1}{4}x + 5$
   B. $y = 4x - 20$
   C. $y = 4x - 5$
   D. $y = 4x + 5$
   E. $y = 5x + 4$

* Recall the **slope-intercept** form for the equation of a line: $y = mx + b$ (see the end of the solution to problem 52).

Perpendicular lines have slopes that are negative reciprocals of each other. Since the slope of the old line is $-\frac{1}{4}$, the slope of the new line is $m = 4$. We are also given that the y-intercept of the new line is $(0,5)$. So $b = 5$. The equation of the line is therefore $y = 4x + 5$, choice **D**.

Solution by plugging in the given point: Since the point $(0,5)$ lies on the line, if we substitute 0 in for x, we should get 5 for y. Let's substitute 0 in for x in each answer choice.

   (A) 5
   (B) –20
   (C) – 5
   (D) 5
   (E) 4

We can eliminate choices B, C and E because they did not come out to 5.

Finally, since the lines are perpendicular, we choose choice **D**.

Recall: Parallel lines have the same slope, and perpendicular lines have slopes that are negative reciprocals of each other.

86. The radius of a circle is $\frac{15}{\pi}$ centimeters. How many centimeters long is its circumference?

   F. $\frac{30}{\pi}$

   G. $\frac{300}{\pi}$

   H. 15

   J. 30

   K. 30π

80

* The **circumference** of a circle is $C = 2\pi r$ where r is the **radius** of the circle. So in this example, the circumference is $C = 2\pi \cdot \dfrac{15}{\pi} = 30$ centimeters, choice **J**.

Definitions: A **circle** is a two-dimensional geometric figure formed of a curved line surrounding a center point, every point of the line being an equal distance from the center point. This distance is called the **radius** of the circle. The **diameter** of a circle is the distance between any two points on the circle that pass through the center of the circle. The perimeter of a circle is called its **circumference** which can be found by using the formula $C = 2\pi r$ where r is the radius of the circle.

Example:

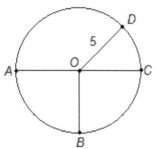

In the figure above we have a circle with center O and radius $r = 5$. Note that $\overline{OA}$, $\overline{OB}$, $\overline{OC}$, and $\overline{OD}$ are all radii of the circle. Also, $\overline{AC}$ is a diameter of the circle with length 10, and the circumference of the circle is $C = 10\pi$.

87. In the figure below, where $\triangle CAT \sim \triangle DOG$, lengths given are in inches. What is the area, in square inches, of $\triangle DOG$?
(Note: The symbol $\sim$ means "is similar to.")

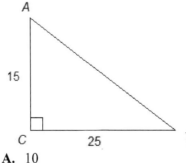

    A. 10
    B. 20
    C. 30
    D. 60
    E. 80

Since corresponding sides of similar triangles are in proportion we can find OD by setting up a ratio, cross multiplying, and dividing as follows: $\frac{OD}{15} = \frac{10}{25}$. So $25OD = 150$, and therefore $OD = \frac{150}{25} = 6$ inches.

Now the area of the triangle is $\frac{1}{2}bh = \frac{1}{2}(10)(6) = 30$ in^2, choice **C**.

Definitions: Two triangles are **similar** if their angles are congruent. Note that similar triangles **do not** have to be the same size. **Corresponding sides of similar triangles are in proportion**.

88. In the figure below, line k is parallel to line n. If line m bisects angle ABC, what is the value of x ?

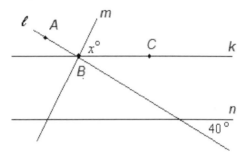

 F. 50
 G. 60
 H. 70
 J. 80
 K. 90

* This is a problem involving two parallel lines cut by a transversal. In this example there are actually two transversals. It is useful to isolate just one of them. We do this below.

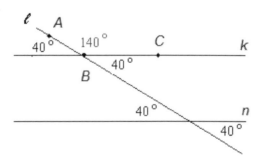

82

Note that the transversal ℓ creates 8 angles, four of which have measure 40 degrees. The other four have measure 140 degrees (only one is labeled in the picture). Any two non-congruent angles are supplementary, that is they add up to 180 degrees. Finally, we note that x is half of 140 because line m bisects angle ABC. Thus $x = 70$, choice **H**.

LEVEL 3: PROBABILITY AND STATISTICS

89. To decrease the mean of 5 numbers by 3, by how much would the sum of the 5 numbers have to decrease?

   A.   3
   B.   5
   C.   7.5
   D.   12
   E.   15

Solution by picking numbers: Let's pick some numbers. The numbers 10, 10, 10, 10, and 10 have a mean of 10 and a sum of 50. The numbers 7, 7, 7, 7, and 7 have a mean of 7 and a sum of 35. So to decrease the mean by 3 we had to decrease the sum by 15, choice **E**.

* Quick solution: One way to decrease the mean of a list of numbers by 3 is to decrease **each** number in the list by 3. Since there are 5 numbers, the sum must be decreased by $3(5) = 15$, choice **E**.

90. A drawer contains 7 white shirts, 6 black shirts, and 5 green shirts. How many black shirts must be added to the drawer so that the probability of randomly choosing a black shirt is $\frac{3}{5}$?

   F.   36
   G.   24
   H.   18
   J.   15
   K.   12

Solution by plugging in answer choices: First note that there are a total of $7 + 6 + 5 = 18$ shirts in the drawer. Also note that as a decimal $\frac{3}{5} = 0.6$.

Let's start with choice H and guess that we need to add 18 black shirts to the drawer. Then there will be $6 + 18 = 24$ black shirts and $18 + 18 = 36$ shirts in total. The probability of randomly choosing a black shirt would then be $\frac{24}{36} \approx 0.67$. This is too big so we can eliminate choices F, G and H.

Let's try choice J next and guess that we need to add 15 black shirts to the drawer. Then there will be $6 + 15 = 21$ black shirts and $18 + 15 = 33$ shirts in total. The probability of randomly choosing a black shirt would then be $\frac{21}{33} \approx 0.63$. This is still too big. So we can eliminate choice J.

The answer must therefore be choice **K**.

As an extra precaution, let's just verify that answer choice K works. If we need to add 12 black shirts to the drawer, then there will be $6 + 12 = 18$ black shirts and $18 + 12 = 30$ shirts in total. The probability of randomly choosing a black shirt would then be $\frac{18}{30} = 0.6$. This is correct. So the answer is in fact choice **K**.

Remarks: (1) To compute a simple probability where all outcomes are equally likely, divide the number of "successes" by the total number of outcomes.

(2) Note that we changed all the fractions to decimals in this problem because decimals are much easier to compare than fractions. This helped us to choose our guesses more efficiently.

* Algebraic solution: There are 6 black shirts and $7 + 6 + 5 = 18$ shirts in total in the drawer. Let x be the number of black shirts we must add to the drawer. Note that the total will also be increased by x. So we get the equation

$$\frac{6+x}{18+x} = \frac{3}{5}$$

We now cross multiply to get $5(6 + x) = 3(18 + x)$. We distribute on both sides of the equation to get $30 + 5x = 54 + 3x$. We now subtract $3x$ from each side of the equation to get $30 + 2x = 54$. We subtract 30 from each side of this equation to get $2x = 24$. Finally we divide each side of this last equation by 2 to get $x = \frac{24}{2} = 12$, choice **K**.

91. Let $p(E)$ denote the probability that event E occurs, and let $p(\sim E)$ denote the probability that event E does not occur. Which of the following statements is **always** true?

   A. $p(E) + p(\sim E) = 1$
   B. $p(E) < p(\sim E)$
   C. $p(E) > p(\sim E)$
   D. $p(E) > 1$
   E. $p(\sim E) > 1$

* If $p(E)$ is the probability that event E occurs, then the probability that event E does not occur is $p(\sim E) = 1 - p(E)$. Adding $p(E)$ to each side of this equation gives $p(E) + p(\sim E) = 1$, choice **A**.

Remarks: (1) Choices D and E are **always** false since probabilities are always between 0 and 1.

(2) Choices B and C can be true or false, depending on the event E.

92. Between Town A and Town B there are 5 roads, between Town B and Town C there are 2 roads, and between Town C and Town D there are 4 roads. If a traveler were to travel from Town A to Town D, passing first through B, then through C, how many different routes does he have to choose from?

   F. 11
   G. 20
   H. 40
   J. 60
   K. 80

* **Solution using the counting principle:** The counting principle says that when you perform events in succession you multiply the number of possibilities. So we get $5 \cdot 2 \cdot 4 = 40$ different routes, choice **H**.

LEVEL 3: TRIGONOMETRY

93. In $\triangle DOG$, the measure of $\angle D$ is 60° and the measure of $\angle O$ is 30°. If $\overline{DO}$ is 8 units long, what is the area, in square units, of $\triangle DOG$?

   A. 4
   B. 8
   C. $8\sqrt{2}$
   D. $8\sqrt{3}$
   E. 16

* **Solution using a 30, 60, 90 right triangle:** Let's draw two pictures.

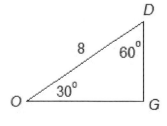
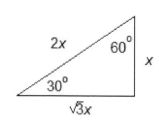

The picture on the left is what is given in the problem. Comparing this to the picture on the right we see that $x = 4$ and $\sqrt{3}x = 4\sqrt{3}$. So the area of the triangle is $\frac{1}{2}(4)(4\sqrt{3}) = 8\sqrt{3}$, choice **D**.

Note: It is worth memorizing the 30, 60, 90 triangle on the right. You should also commit the 45, 45, 90 triangle to memory (see the end of the solution to problem 62 for details).

Trigonometric solution: We have $\sin 30° = \frac{DG}{8}$. So $DG = 8\sin 30°$. Similarly, $\cos 30° = \frac{OG}{8}$. So $OG = 8\cos 30°$. So the area of the triangle is $\frac{1}{2}(OG)(DG) = \frac{1}{2}(8\cos 30°)(8\sin 30°) \approx 13.8564$ (using our calculator).

Now plug the answer choices into the calculator and we see that $8\sqrt{3} \approx 13.8564$ choice **D**.

Remark: Make sure that your calculator is in degree mode before using your calculator. Otherwise you will get the wrong answer when estimating values of trig functions.

If you are using a TI-84 (or equivalent) calculator press MODE and on the third line, make sure that DEGREE is highlighted. If it is not, scroll down and select it. If possible, do not alter this setting until you are finished taking your ACT.

94. In right triangle PQR below, the measure of $\angle R$ is 90°, $PQ = 4$ units, and $RQ = 3$ units. What is $\tan Q$?

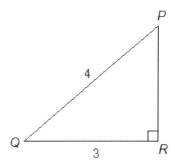

F. $\dfrac{2}{3}$

G. $\dfrac{3}{4}$

H. $\dfrac{\sqrt{7}}{3}$

J. $\dfrac{3}{\sqrt{7}}$

K. $\dfrac{4}{3}$

* We first find PR by using the Pythagorean Theorem: $3^2 + PR^2 = 4^2$. So $9 + PR^2 = 16$. Subtracting 9 from each side gives us $PR^2 = 16 - 9 = 7$, and therefore $PR = \sqrt{7}$.

Finally, we have $\tan Q = \dfrac{\text{OPP}}{\text{ADJ}} = \dfrac{\sqrt{7}}{3}$, choice **H**.

Remarks: (1) The Pythagorean Theorem says that if a right triangle has legs of length a and b, and a hypotenuse of length c, then $c^2 = a^2 + b^2$.

In this example note that PR is one of the legs and **not** the hypotenuse.

(2) If you do not see why we have $\tan Q = \dfrac{\text{OPP}}{\text{ADJ}}$, review the basic trigonometry given after the solution to problem 61.

95. A dog, a cat, and a mouse are all sitting in a room. Their relative positions to each other are described in the figure below. Which of the following expressions gives the distance, in feet, from the cat to the mouse?

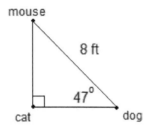

A. $8 \sin 47°$

B. $8 \cos 47°$

C. $8 \tan 47°$

D. $\dfrac{8}{\sin 47°}$

E. $\dfrac{8}{\cos 47°}$

* Let x be the distance, in feet, from the cat to the mouse. Then we have $\sin 47° = \dfrac{\text{OPP}}{\text{HYP}} = \dfrac{x}{8}$. Multiplying each side of the equation $\sin 47° = \dfrac{x}{8}$ by 8 gives us $8\sin 47° = x$. So the answer is choice **A**.

Remarks: (1) If you do not see why we have $\sin 47° = \dfrac{\text{OPP}}{\text{HYP}}$, review the basic trigonometry given after the solution to problem 61.

(2) If you prefer, you can think of the multiplication above as **cross multiplication** by first rewriting $\sin 47°$ as $\frac{\sin 47°}{1}$.

So we have $\frac{\sin 47°}{1} = \frac{x}{8}$. Cross multiplying yields $8\sin 47° = 1x$. This yields the same result as in the above solution.

96. In the figure below, $\sin k = ?$

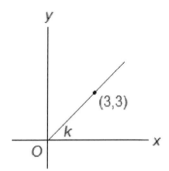

    F. $3\sqrt{2}$

    G. $\frac{\sqrt{2}}{2}$

    H. $\frac{\sqrt{3}}{2}$

    J. 1

    K. $\frac{1}{3}$

* Let's add a little information to the picture.

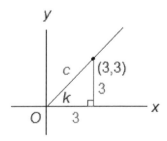

By the Pythagorean Theorem, $c^2 = 3^2 + 3^2 = 9 + 9 = 18$. So $c = \sqrt{18}$.

Now, $\sin k = \dfrac{OPP}{HYP} = \dfrac{3}{\sqrt{18}} \sim .707$. We now use our calculator to approximate each answer choice, and we see that $\dfrac{\sqrt{2}}{2} \sim .707$. So the answer is choice **G**.

Remarks: (1) $\sqrt{18}$ can be simplified as $\sqrt{18} = \sqrt{9 \cdot 2} = \sqrt{9}\sqrt{2} = 3\sqrt{2}$. So $\dfrac{3}{\sqrt{18}} = \dfrac{3}{3\sqrt{2}} = \dfrac{1}{\sqrt{2}}$.

Furthermore, we can rationalize the denominator in the expression $\dfrac{1}{\sqrt{2}}$ to get $\dfrac{1}{\sqrt{2}} \cdot \dfrac{\sqrt{2}}{\sqrt{2}} = \dfrac{\sqrt{2}}{2}$. This is a way to see that the answer is choice G without using our calculator.

(2) Instead of using the Pythagorean Theorem, we can observe that the triangle we formed is an isosceles right triangle which is the same as a 45, 45, 90 right triangle. So the hypotenuse of the triangle has length $3\sqrt{2}$ (see the end of the solution to problem 62 for details).

LEVEL 4: NUMBER THEORY

97. What is the value of $\log_4 64$?

    A. 3
    B. 4
    C. 6
    D. 10
    E. 16

Calculator solution by changing the base: Your calculator can only compute logarithms with a base of 10 (or the number e). We use the change of base formula and our calculator to get

$$\log_4 64 = \frac{\log 64}{\log 4} = 3$$

This is choice **A**.

Note: (1) $\log x$ means $\log_{10} x$

(2) To change any logarithm to base 10 use the formula $\log_b a = \dfrac{\log a}{\log b}$.

Straightforward calculator solution: If you have the latest TI-84 Plus calculator you can do this computation directly as follows:

Press ALPHA, then Y=, then WINDOW, and then 5 to select logBASE(.

You can now enter the 4 and 64 in the appropriate positions and hit ENTER to get $\log_4 64 = 3$, choice **A**.

Solution by changing to exponential form: The logarithmic equation $y = \log_4 64$ is equivalent to the exponential equation $4^y = 64$. Since 64 can be written as 4^3, we have $4^y = 4^3$. Therefore, $y = 3$, choice **A**.

Notes: (1) The word "logarithm" just means "exponent."

(2) The equation $y = \log_4 64$ can be read as "y is the exponent when we rewrite 64 with a base of 4." So y is the exponent and 4 is the base. In other words, we are raising 4 to the power y. So we have $4^y = 64$.

(3) When we have an equation involving exponents, if the bases on each side of the equation are equal, then so are the exponents. So once we have $4^y = 4^3$, we can say $y = 3$.

* **Quick solution:** The question is asking "4 to what power is 64?" Well 4 to the third power is 64, so the answer is 3, choice **A**.

98. In a game of catch, 5 people stand in a circle. One person (the thrower) throws the ball to another person (the catcher). The thrower cannot throw the ball to the person immediately to their left or right, or to the person who last threw the ball. When will the first person who throws the ball become the catcher for the first time?

    F.   3rd
    G.   4th
    H.   5th
    J.   6th
    K.   10th

* Let's draw a picture.

Note that the first thrower has two choices, but after this the picture is determined. In either case, the first thrower (labeled "Start" in the picture) becomes the first catcher 5th, choice **H**.

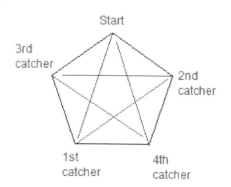

99. If b is a positive integer that divides both 98 and 140, but divides neither 20 nor 35, what should you get when you add the digits in b ?

  A. 1
  B. 3
  C. 4
  D. 5
  E. 6

* We start by finding the **greatest common divisor (gcd)** of 98 and 140. On your TI-84 calculator press MATH, scroll right to NUM. For the gcd press 9, type 98, 140 and press ENTER. You will see an output of 14. Now note that 14 does not divide 20 or 35. So $b = 14$ and when we add the digits in b we get $1 + 4 = 5$, choice **D**.

Remarks: (1) See problem 39 for more information on the gcd and for other ways to solve this problem.

(2) Once we have the gcd of two numbers, we can find all the positive divisors of those two numbers by taking all positive divisors of the gcd. So the list of all positive integers that divide 98 and 140 are 1, 2, 7, and 14. Note that 1 and 2 divide 20 and 7 divides 35 so that b cannot be equal to 1, 2, or 7.

100. Two numbers are reciprocals if their product is equal to 1. If a and b are reciprocals and $a < -1$, then b must be:

  F. greater than 1
  G. between 0 and 1
  H. equal to 0
  J. between -1 and 0
  K. less than -1

* **Solution by picking a number:** Let's choose a value for a, say $a = -2$. Then $b = -\frac{1}{2}$. This number is between -1 and 0, choice **J**.

Warning: Do not confuse "reciprocal" with "negative reciprocal." To find the reciprocal we just flip the fraction, so that the reciprocal of $\frac{a}{b}$ is $\frac{b}{a}$. Equivalently, we can put 1 over the number. So the reciprocal of x is $\frac{1}{x}$.

101. The ratio of x to y is 5 to 1, and the ratio of y to z is 1 to 4. What is the value of $\frac{3x-2y}{5y+2z}$?

    A. 1
    B. 2
    C. 3
    D. 4
    E. 5

* Solution by picking numbers: Let's choose values for x, y, and z that satisfy the given condition, say $x = 5$, $y = 1$, and $z = 4$. Then we have

$$\frac{3x-2y}{5y+2z} = \frac{3(5)-2(1)}{5(1)+2(4)} = \frac{15-2}{5+8} = \frac{13}{13} = 1,$$ choice **A.**

Complete algebraic solution: We are given that $\frac{x}{y} = \frac{5}{1}$ and $\frac{y}{z} = \frac{1}{4}$. Cross multiplying these two equations gives $x = 5y$ and $z = 4y$. We now substitute for x and z in the given expression:

$$\frac{3x-2y}{5y+2z} = \frac{3(5y)-2y}{5y+2(4y)} = \frac{15y-2y}{5y+8y} = \frac{13y}{13y} = 1$$

This is choice **A.**

102. The decimal representation of $\frac{11}{13}$ repeats and can be written as 0.846153846153... What is the 200th digit to the right of the decimal point in this decimal representation?

    F. 1
    G. 3
    H. 4
    J. 5
    K. 6

* Since there are exactly 6 digits before repeating we look for the remainder when 200 is divided by 6. One way to do this is to first find an integer as close to 200 as possible that is divisible by 6. We check this in our calculator.

$$200/6 \approx 33.33$$
$$199/6 \approx 33.17$$
$$198/6 = 33$$

So 198 is divisible by 6 and therefore 200 gives a remainder of 2 when divided by 6. So the digit in the 200th place is the same as the digit in the second place to the right of the decimal point. This is 4, choice **H.**

Remark: Instead of guessing with your calculator you can also divide 200 by 6 using long division to find the remainder. Note that it is **not** correct to simply divide 200 by 6 in your calculator. As you can see above you would get approximately 33.33 which has nothing to do with the remainder (which is 2).

Calculator Algorithm for computing a remainder: Although performing division in your calculator never produces a remainder, there is a simple algorithm you can perform which mimics long division. Let's find the remainder when 200 is divided by 6 using this algorithm.

Step 1: Perform the division in your calculator: $200/6 \sim 33.33$
Step 2: Multiply the integer part of this answer by the divisor: $33*6 = 198$
Step 3: Subtract this result from the dividend to get the remainder:

$$200 - 198 = \mathbf{2}$$

103. $\dfrac{1}{4} \cdot \dfrac{3}{5} \cdot \dfrac{4}{6} \cdot \dfrac{5}{7} \cdot \dfrac{6}{8} \cdot \dfrac{7}{9} \cdot \dfrac{8}{10} \cdot \dfrac{9}{11} \cdot \dfrac{10}{12} \cdot \dfrac{11}{13} = ?$

    A. $\dfrac{1}{156}$

    B. $\dfrac{1}{52}$

    C. $\dfrac{1}{4}$

    D. 1

    E. $\dfrac{50}{3}$

* Note that we have one large product here. So we can cancel a number in any denominator with the same number in any numerator. For example, the 4 in the denominator of the first fraction can be canceled with the 4 in the numerator of the third fraction. Continuing this way, we cancel the 5's, 6's and so on through the 11's.

$$\dfrac{1}{4} \cdot \dfrac{3}{5} \cdot \dfrac{4}{6} \cdot \dfrac{5}{7} \cdot \dfrac{6}{8} \cdot \dfrac{7}{9} \cdot \dfrac{8}{10} \cdot \dfrac{9}{11} \cdot \dfrac{10}{12} \cdot \dfrac{11}{13} = \dfrac{1 \cdot 3}{12 \cdot 13} = \dfrac{1}{52}$$

This is choice **B**.

104. On Monday, Daniel received m dollars for his birthday and spent $\frac{1}{3}$ of it. On Tuesday, Daniel spent $\frac{1}{3}$ of the amount that he had left. He repeated this on Wednesday and Thursday, spending $\frac{1}{3}$ of the amount he had left each time. On Friday he spent the remaining n dollars. If m and n are both integers, what is the least possible value for m ?

    F. 81
    G. 54
    H. 27
    J. 15
    K. 12

Solution by plugging in answer choices: Since the word "least" appears in the problem, let's start with the smallest answer choice, choice K. If we guess that Daniel received 12 dollars, then he spent $\frac{1}{3}(12) = 4$ dollars on Monday, leaving him with $12 - 4 = 8$ dollars. Since 8 is not divisible by 3 we can eliminate choice K.

Let's try choice J next and guess that Daniel received 15 dollars. Then he spent $\frac{1}{3}(15) = 5$ dollars on Monday, leaving him with $15 - 5 = 10$ dollars. Since 10 is not divisible by 3 we can eliminate choice J.

Let's try H next and guess that Daniel received 27 dollars. Then he spent $\frac{1}{3}(27) = 9$ dollars on Monday, leaving him with $27 - 9 = 18$ dollars. He then spent $\frac{1}{3}(18) = 6$ dollars on Tuesday, leaving him $18 - 6 = 12$ dollars. On Wednesday he spent $\frac{1}{3}(12) = 4$ dollars, leaving him $12 - 4 = 8$ dollars. Since 8 is not divisible by 3 we can eliminate choice H.

Let's try choice G next and guess that Daniel received 54 dollars. Then he spent $\frac{1}{3}(54) = 18$ dollars on Monday, leaving him with $54 - 18 = 36$ dollars. He then spent $\frac{1}{3}(36) = 12$ dollars on Tuesday, leaving him with $36 - 12 = 24$ dollars. On Wednesday he spent $\frac{1}{3}(24) = 8$ dollars, leaving him with $24 - 8 = 16$ dollars. Since 16 is not divisible by 3 we can eliminate choice G.

94

So the answer must be choice F. But let's check this to be sure. If Daniel received 81 dollars, then he spent $\frac{1}{3}(81) = 27$ dollars on Monday, leaving him with $81 - 27 = 54$ dollars. He then spent $\frac{1}{3}(54) = 18$ dollars on Tuesday, leaving him with $54 - 18 = 36$ dollars. On Wednesday he spent $\frac{1}{3}(36) = 12$ dollars, leaving him with $36 - 12 = 24$ dollars. On Thursday he spent $\frac{1}{3}(24) = 8$ dollars, leaving him with $24 - 8 = 16$ dollars. And finally on Friday he spends the last $n = 16$ dollars. So the answer is in fact **F**.

* **Quick solution:** Each day we are taking $\frac{1}{3}$ of the previous number and then subtracting this from the original number. This is equivalent to taking $\frac{2}{3}$ of the number. So we have $n = \frac{2}{3} \cdot \frac{2}{3} \cdot \frac{2}{3} \cdot \frac{2}{3} \cdot m = \frac{16}{81}m$. So the answer must be divisible by 81. Of the answers listed, only 81 satisfies this condition. So the answer is choice **F**.

LEVEL 4: ALGEBRA AND FUNCTIONS

105. The *determinant* of a matrix $\begin{bmatrix} a & b \\ c & d \end{bmatrix}$ is equal to $ad - bc$. What must be the value of z for the matrix $\begin{bmatrix} z & z \\ z & 6 \end{bmatrix}$ to have a determinant of 9 ?

    A. −6
    B. −3
    C. $-\dfrac{12}{5}$
    D. $\dfrac{12}{7}$
    E. 3

$\begin{bmatrix} z & z \\ z & 6 \end{bmatrix} = (z)(6) - (z)(z) = 6z - z^2$. We can now proceed in two ways.

Solution by plugging in answer choices: Normally we would start with choice C, but in this case there is no real advantage to doing so. We might as well start with choice E since it is the easiest to plug in. If $z = 3$, then $6z - z^2 = 6(3) - 3^2 = 18 - 9 = 9$. This is correct. So the answer is choice **E**.

* **Algebraic solution:** We need to solve the equation $6z - z^2 = 9$. Bringing everything over to the right gives $0 = z^2 - 6z + 9$. We can factor on the right to get $0 = (z - 3)^2$. So $z - 3 = 0$ and so $z = 3$, choice **E**.

106. What is the sum of the 2 solutions of the equation $x^2 - 7x + 3$?

    F. -7
    G. -3
    H. 0
    J. 3
    K. 7

* **Quick solution:** The sum of the solutions is the negative of the coefficient of the x term. So the answer is 7, choice **K.**

Notes: (1) If r and s are the solutions of the quadratic equation $x^2 + bx + c = 0$, then $b = -(r + s)$ and $c = rs$. So in this problem the sum of the 2 solutions is 7 and the product of the 2 solutions is 3.

(2) Yes, you can also solve the equation $x^2 - 7x + 3 = 0$ by completing the square or using the quadratic formula, but this is very time consuming. It is much better to use the quick solution given above.

107. In the real numbers, what is the solution of the equation $4^{x+2} = 8^{2x-1}$?

    A. $-\dfrac{7}{4}$

    B. $-\dfrac{1}{4}$

    C. $\dfrac{3}{4}$

    D. $\dfrac{5}{4}$

    E. $\dfrac{7}{4}$

* **Algebraic solution:** The numbers 4 and 8 have a common base of 2. In fact, $4 = 2^2$ and $8 = 2^3$. So we have $4^{x+2} = (2^2)^{x+2} = 2^{2x+4}$ and we have $8^{2x-1} = (2^3)^{2x-1} = 2^{6x-3}$. Thus, $2^{2x+4} = 2^{6x-3}$. So $2x + 4 = 6x - 3$. We subtract $2x$ from each side of this equation to get $4 = 4x - 3$. We now add 3 to each side of this last equation to get $7 = 4x$. Finally we divide each side of this equation by 4 to get $\dfrac{7}{4} = x$, choice **E.**

Notes: (1) For a review of the laws of exponents used here see the end of the solution to problem 43.

96

(2) This problem can also be solved by plugging in (start with choice C). We leave it to the reader to solve the problem this way. Make sure to use your calculator.

108. For all nonzero j and k, $\dfrac{(j \times 10^3)(k^2 \times 0.001)}{(j^2 \times 10,000)(k \times 10^{-3})} = ?$

    F. $\dfrac{1}{10}$

    G. $\dfrac{k}{j}$

    H. $\dfrac{k}{j}$

    J. $\dfrac{k}{10j}$

    K. $\dfrac{k^2}{10j^2}$

$*$ $\dfrac{(j \times 10^3)(k^2 \times 0.001)}{(j^2 \times 10,000)(k \times 10^{-3})} = \dfrac{j}{j^2} \times \dfrac{k^2}{k} \times \dfrac{10^3}{10,000} \times \dfrac{0.001}{10^{-3}} = \dfrac{1}{j} \times \dfrac{k}{1} \times \dfrac{1}{10} \times 1 = \dfrac{k}{10j}$

This is choice **J**.

Notes: (1) $10^3 = 1000$ so that $\dfrac{10^3}{10,000} = \dfrac{1000}{10,000} = \dfrac{1}{10}$.

(2) $10^{-3} = 0.001$ so that $\dfrac{0.001}{10^{-3}} = \dfrac{0.001}{0.001} = 1$.

109. After solving a quadratic equation by completing the square, it was found that the equation had solutions, $x = -3 \pm \sqrt{-16b^2}$ where b is a positive real number. Which of the following gives the solutions as complex numbers?

    A. $-3 \pm \quad bi$
    B. $-3 \pm \quad 4bi$
    C. $-3 \pm \quad 8bi$
    D. $-3 \pm 16bi$
    E. $-3 \pm 256bi$

Since both the question and answer choices all begin with -3, it suffices to evaluate $\sqrt{-16b^2}$.

Since $4^2 = 16$, choice B looks like it should be the answer. Indeed we have $(4bi)^2 = (4bi)(4bi) = 16b^2 i^2 = 16b^2(-1) = -16b^2$. It follows that $\sqrt{-16b^2} = 4bi$, and so the answer is choice **B**.

* Quick solution: $\sqrt{-16b^2} = \sqrt{-16}\sqrt{b^2} = 4i \cdot b = 4bi$, choice **B**.

Technical Remark: If a and b are nonnegative real numbers, then it is always true that $\sqrt{ab} = \sqrt{a}\sqrt{b}$, but this rule is NOT always true when dealing with complex numbers.

For example, if a and b are both -2, then $\sqrt{ab} = \sqrt{(-2)(-2)} = \sqrt{4} = 2$, but $\sqrt{a}\sqrt{b} = \sqrt{-2}\sqrt{-2} = \sqrt{2}i\sqrt{2}i = 2i^2 = -2$.

So we need to be extra careful with the above reasoning. In this problem we do get the correct answer.

110. Consider the rational function $r(x) = \dfrac{x^2-5}{x-3}$. Let $m = r(5)$, let n be the number of horizontal and/or vertical asymptotes there are for the graph of r. What is the value of $m \cdot n$?

    F. 30
    G. 20
    H. 10
    J. 5
    K. 0

* $m = r(5) = \dfrac{5^2-5}{5-3} = \dfrac{25-5}{2} = \dfrac{20}{2} = 10$.

The graph also has a vertical asymptote of $x = 3$. There are no horizontal asymptotes. So $n = 1$.

So $m \cdot n = 10(1) = 10$, choice **H**.

Notes: (1) The variable x is a placeholder. We evaluate the function r at a specific value by substituting that value in for x. In this question we replaced x by 5.

(2) To find vertical asymptotes of a rational function on the ACT simply find all numbers that will make the denominator of the function 0. If plugging in the number a makes the denominator 0, then the vertical line $x = a$ is a vertical asymptote for the function. You can usually find these numbers by simple observation, but if necessary you can set the denominator equal to zero and formally solve the resulting equation.

In this problem it is easy to see that plugging a 3 into the function makes the denominator 0, so $x = 3$ is a vertical asymptote.

(3) In this example the polynomial in the numerator has a higher degree than the polynomial in the denominator. It follows that there are no horizontal asymptotes.

(4) A complete discussion of vertical and horizontal asymptotes is outside of the scope of this book. Here I am providing only what is necessary to answer ACT problems correctly.

As an example, in general, if the number a makes the denominator of a rational function 0 there is no guarantee that $x = a$ is a vertical asymptote. If the number a also makes the numerator 0, then $x = a$ may or may not be a vertical asymptote.

111. For all real numbers x and y, $|x - y|$ is equivalent to which of the following?

　　A.　$x + y$
　　B.　$\sqrt{x - y}$
　　C.　$(x - y)^2$
　　D.　$\sqrt{(x - y)^2}$
　　E.　$-(x - y)$

* Solution using the definition of absolute value: One definition of the absolute value of x is $|x| = \sqrt{x^2}$. So $|x - y| = \sqrt{(x - y)^2}$, choice **D.**

Note: Here we have simply replaced x by $x - y$ on both sides of the equation $|x| = \sqrt{x^2}$.

Solution by picking numbers: Let's choose values for x and y, let's say $x = 2$ and $y = 5$. Then $|x - y| = |2 - 5| = |{-3}| = \mathbf{3}$.

Put a nice big dark circle around **3** so you can find it easily later. We now substitute $x = 2$ and $y = 5$ into each answer choice:

　　A.　7
　　B.　$\sqrt{-3}$
　　C.　$(-3)^2 = 9$
　　D.　$\sqrt{(-3)^2} = \sqrt{9} = 3$
　　E.　$-(-3) = 3$.

Since A, B and C each came out incorrect, we can eliminate them. But we still do not know if the answer is D or E. We will have to pick new numbers.

Let's try $x = 5$ and $y = 2$ this time. Then $|x - y| = |5 - 2| = |3| = \mathbf{3}$.

Put a nice big dark circle around **3** so you can find it easily later. We now substitute $x = 5$ and $y = 2$ into each of the remaining answer choices:

> **D.** $\sqrt{3^2} = \sqrt{9} = 3$
>
> **E.** -3.

Since E came out incorrect, the answer is choice **D.**

112. A function h is defined as follows:
$$\text{for } x > 0, h(x) = x^7 + 2x^5 - 12x^3 + 15x - 2$$
$$\text{for } x \leq 0, h(x) = x^6 - 3x^4 + 2x^2 - 7x - 5$$
What is the value of $h(-1)$?

> **F.** -8
> **G.** -4
> **H.** 0
> **J.** 2
> **K.** 4

* Since $-1 \leq 0$, we use the second equation. It follows that
$h(-1) = (-1)^6 - 3(-1)^4 + 2(-1)^2 - 7(-1) - 5 = 1 - 3 + 2 + 7 - 5 = 2$.
This is choice **J.**

LEVEL 4: GEOMETRY

113. For what value of k would the following system of equations have an infinite number of solutions?
$$5x - 2y = 9$$
$$20x - 8y = 9k$$

> **A.** 2
> **B.** 4
> **C.** 8
> **D.** 27
> **E.** 36

The system of equations

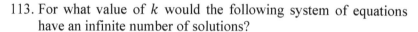

$$ax + by = c$$
$$dx + ey = f$$

100

has an infinite number of solutions if $\frac{a}{d} = \frac{b}{e} = \frac{c}{f}$. So we solve the equation $\frac{5}{20} = \frac{9}{9k}$. Cross multiplying yields $45k = 180$ so $k = \frac{180}{45} = 4$, choice **B**.

Note: In this problem $\frac{b}{e} = \frac{a}{d}$. Indeed, $\frac{5}{20} = \frac{-2}{-8}$. This guarantees that the system of equations has infinitely many solutions instead of no solutions.

* **Quick solution:** We multiply 5 by 4 to get 20 (or –2 by 4 to get –8). So $9k = 4(9) = 36$, and therefore $k = \frac{36}{9} = 4$, choice **B**.

See the end of problem 83 for further explanation.

114. A right circular cylinder has a base diameter and height of 10 inches each. What is the total surface area of this cylinder, in square inches?

    F. 25π
    G. 50π
    H. 75π
    J. 150π
    K. 200π

* When we cut and unfold the cylinder we get the following rectangle.

10

$C = 10\pi$

Notice that the width of the rectangle is the circumference of the base of the cylinder. Thus the width is

$$C = \pi d = \pi(10) = 10\pi \text{ inches.}$$

The **lateral** surface area of the cylinder is the area of this rectangle.

$$L = 10(10\pi) = 100\pi \text{ square inches.}$$

We also need the area of the two bases. Each of these is a circle with area $A = \pi r^2 = \pi(5)^2 = 25\pi$ inches. Therefore, the total surface area is

$$S = L + 2A = 100\pi + 2(25\pi) = 150\pi \text{ square inches.}$$

This is choice **J**.

101

Note: The radius of a circle is half of the diameter. So in this problem the radius of a base of the cylinder is $r = \dfrac{d}{2} = \dfrac{10}{2} = 5$.

115. Three distinct lines, all contained in a plane, separate the plane into distinct regions. Which of the following is a possibility for the number of distinct regions of the plane that may be separated by any 3 such lines?

    A. 2
    B. 3
    C. 5
    D. 6
    E. 8

* Here are two different pictures which lead to one of the answer choices:

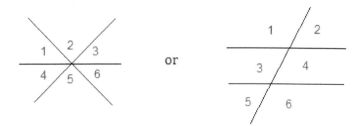

We see from either of the pictures above that it is possible for there to be 6 regions, choice **D**.

Remark: The figures above give pictures when there are 1 (left) or 2 (right) points of intersection among the lines. There can also be 0 or 3 points of intersection leading to 4 or 7 regions, respectively. Here are pictures of those two possibilities.

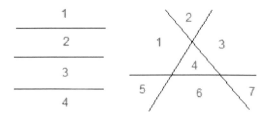

116. In the (x, y) coordinate plane, what is the radius of the circle having the points $(2, -4)$ and $(-4, 4)$ as endpoints of a diameter?

    F.   5
    G. 10
    H. 16
    J. 25
    K. 100

Solution using a right triangle: Let's plot the two points and form a right triangle.

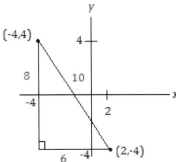

We got the length of the left leg by subtracting $4 - (-4) = 4 + 4 = 8$, and we got the bottom leg by subtracting $2 - (-4) = 2 + 4 = 6$. We now use the Pythagorean Theorem: $c^2 = 6^2 + 8^2 = 36 + 64 = 100$. So $c = 10$.

It follows that the diameter of the circle is 10, and therefore the radius of the circle is 5, choice **F**.

Remarks: (1) If you recognize that 6, 8, 10 is a multiple of the **Pythagorean triple** 3, 4, 5 (just multiply each number by 2), then you do not need to use the Pythagorean Theorem.

(2) 3, 4, 5 and 5, 12, 13 are the two most common Pythagorean triples.

(3) The radius of a circle is $\frac{1}{2}$ the diameter, or $r = \frac{1}{2}d$.

(4) Here is a picture of the circle, the two points, and the diameter.

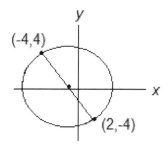

* Solution using the distance formula: We can find the length of the diameter of the circle by using the distance formula. We have

$$d = \sqrt{(-4 - 2)^2 + \left(4 - (-4)\right)^2} = \sqrt{(-6)^2 + 8^2} = \sqrt{36 + 64} = \sqrt{100} = 10$$

It follows that the radius is $\frac{10}{2} = 5$, choice **F.**

Note: The distance between the two points (a,b) and (c,d) is given by

$$d = \sqrt{(c - a)^2 + (d - b)^2} \quad \text{or equivalently} \quad d^2 = (c - a)^2 + (d - b)^2$$

This formula is called the **distance formula**.

117. In pentagon *HOUSE* below, $\angle H$ measures 40°. What is the sum of the measures of the other 4 angles?

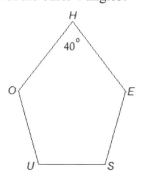

A. 100°
B. 200°
C. 300°
D. 400°
E. 500°

* Solution using a formula: The total number of degrees in the interior of an n-sided polygon is $(n - 2) \cdot 180$. In this problem $n = 5$ so that the total number of degrees is $3(180) = 540$. So the sum of the measures of the <u>other</u> 4 angles is $540 - 40 = 500°$, choice **E.**

104

Solution by drawing a picture: We split the figure up into a triangle and quadrilateral.

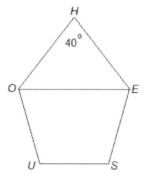

Now recall that a triangle has 180° and a quadrilateral has 360°. It follows that a pentagon has $180 + 360 = 540°$ in total. So the sum of the measures of the <u>other</u> 4 angles is $540 - 40 = 500°$, choice **E**.

118. In the figure below, the circle with center O has a radius of 6 centimeters and the measure of arc PQ is 50°. What is the measure of $\angle QPO$?

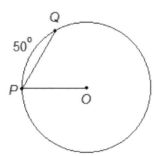

    F.   45°
    G.   65°
    H.   75°
    J.   100°
    K.   130°

* Solution using an inscribed angle: Let's add a bit of information to the picture.

105

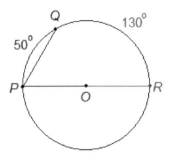

Note that arc QR has $180 - 50 = 130°$. Since $\angle QPO = \angle QPR$ is an **inscribed** angle, its degree measure is half the degree measure of the arc it intercepts. So the measure of $\angle QPO$ is $\frac{130}{2} = 65°$, choice **G.**

Solution using an isosceles triangle: Once again let's add a bit of information to the picture.

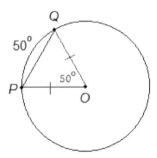

Since $\angle POQ$ is a **central** angle, its degree measure is equal to the degree measure of the arc it intercepts. So the measure of $m\angle POQ = 50°$.

Note that $OP = OQ$ because they are both radii of the circle, and therefore an isosceles triangle is formed. It follows that both base angles are equal in measure.

So each base angle has measure $\frac{180-50}{2} = \frac{130}{2} = 65°$, choice **G.**

119. The area of a rectangular garden is 448 square feet. A statue with a rectangular base that is twice as long as it is wide is placed in the center of the garden so that the garden extends 6 feet beyond the base of the statue on the top and bottom and 4 feet beyond the base of the statue on the left and right sides. How many feet wide is the base of the statue?

    A.   6
    B.   8
    C. 14
    D. 20
    E. 28

Let's draw a picture.

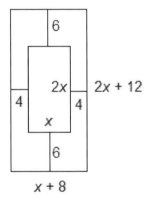

In this picture the larger rectangle represents the garden and the smaller rectangle inside represents the statue. If we let x be the width of the statue, then the length is $2x$ (since the statue is twice as long as it is wide). Since the garden extends 6 feet beyond the statue on top and bottom, the length of the garden is $2x + 6 + 6 = 2x + 12$ feet. Since the garden extends 4 feet beyond the statue on the left and the right, the width of the garden is $x + 4 + 4 = x + 8$ feet.

It follows that the area of the garden is $(x + 8)(2x + 12)$. We need to find the value of x that makes this expression 448. Here are two ways we can do this:

* Method 1 – Plugging in answer choices: Let's start with choice C and guess that $x = 14$. It follows that

$$(x + 8)(2x + 12) = (14 + 8)(2 \cdot 14 + 12) = 22 \cdot 40 = 880.$$

This is too big so we can eliminate choices C, D and E.

Let's try choice B next and guess that $x = 8$. It follows that

$$(x + 8)(2x + 12) = (8 + 8)(2 \cdot 8 + 12) = 16 \cdot 28 = 448.$$

This is correct. So the answer is choice **B**.

Method 2 – Solving a quadratic equation: We need to solve the equation $(x + 8)(2x + 12) = 448$. Multiplying out the left hand side gives $2x^2 + 12x + 16x + 96 = 448$. Dividing by 2 gives $x^2 + 6x + 8x + 48 = 224$. We now combine like terms on the left to get $x^2 + 14x + 48 = 224$. Subtract 224 from each side and we have $x^2 + 14x - 176 = 0$. The left hand side can be factored to get $(x - 8)(x + 22) = 0$. So we have $x - 8 = 0$ or $x + 22 = 0$. Therefore, $x = 8$ or $x = -22$. We reject the negative solution to get $x = 8$, choice **B**.

Notes: (1) As you can see from method 2, a complete algebraic solution is quite tedious here. Plugging in is a better choice for most students.

(2) We can also plug in answer choices right at the beginning of the problem before doing any algebra at all.

(3) The equation $x^2 + 14x - 176 = 0$ can be solved several different ways. We did it by factoring above, but completing the square and using the quadratic formula are two other alternatives.

120. Which of the following expresses the number of feet a contestant must travel in a 5-lap race where the course is a circle of area πR^2 square feet?

    F. $5R$
    G. $5\pi R$
    H. $5\pi R^2$
    J. $10R$
    K. $10\pi R$

A circle with area πR^2 square feet has a radius of R feet, and therefore a circumference of $2\pi R$ feet. So the contestant must travel $2\pi R$ feet per lap. To get the distance needed to travel for 5 laps, we simply multiply by 5 to get $5(2\pi R) = 10\pi R$ feet, choice **K**.

* An even quicker solution: A circle with area πR^2 has circumference $2\pi R$ so the answer is $5(2\pi R) = 10\pi R$, choice **K**.

LEVEL 4: PROBABILITY AND STATISTICS

121. An integer from 50 through 699, inclusive, is to be chosen at random. What is the probability that the number chosen will have 2 as at least one digit?

   A. $\frac{19}{650}$

   B. $\frac{200}{651}$

   C. $\frac{4}{13}$

   D. $\frac{201}{649}$

   E. $\frac{200}{649}$

* From 50 through 699 there are $699 - 50 + 1 = 650$ integers (see Note (3) below).

Now let's count the integers from 50 through 699 that have 2 as at least one digit. We will do this by listing these integers until we recognize a pattern. Note especially the numbers in bold (these are the numbers that many students miscount).

52, 62, 72, 82, 92   **(5 integers)**

102, 112, **120, 121, 122, 123, 124, 125, 126, 127, 128, 129**, 132, 142, 152, 162, 172, 182, 192   **(19 integers)**

200, 201, 202,…, 299   **(100 integers)**

302, 312, 320, 321,…, 392   **(19 integers)**

402, 412, 420, 421,…, 492   **(19 integers)**

502, 512, 520, 521,…592   **(19 integers)**

602, 612, 620, 621,…, 692   **(19 integers)**

The total is $5 + 100 + 5(19) = 200$.

So the probability is $\frac{200}{650} = \frac{4}{13}$, choice **C**.

Notes: (1) To compute a simple probability where all outcomes are equally likely, divide the number of "successes" by the total number of outcomes.

109

In this problem, the total is 650 integers, and 200 are "successes."

(2) The word **inclusive** means that the endpoints are included. So in this example we should include 50 and 699 when computing the total number of outcomes.

(3) To compute the total we used the **Fence-post formula** which says that the number of integers from a to b, inclusive is $b - a + 1$.

To convince you that this formula is correct, let's look at a simple example.

Let's count the number of integers from 5 to 12, inclusive. They are 5, 6, 7, 8, 9, 10, 11, 12, and we see that there are 8 of them. Now $12 - 5 = 7$ which is not the correct amount, but $12 - 5 + 1 = 8$ which is the correct amount.

(4) The list written out above may be much more than is needed (it depends on the individual student). You should keep listing integers until you are convinced that you have a clear picture of the pattern and will not make a careless computational error.

(5) You can quickly reduce a fraction in your TI-84 calculator by performing the division and then pressing MATH ENTER ENTER. In this case we type 200 / 650 MATH ENTER ENTER.

122. An urn contains 15 blue marbles, 7 white marbles, and 8 black marbles. How many additional black marbles must be added to the urn so that the probability of randomly drawing a black marble is $\frac{5}{6}$?

   F.   5
   G.   23
   H.   51
   J.   100
   K.   102

Solution by plugging in answer choices: Let's start with choice H and guess that 51 black marbles must be added to the urn. Then there will be $8 + 51 = 59$ black marbles and $30 + 51 = 81$ marbles in total. So the probability of drawing a black marble will be $\frac{59}{81} \approx 0.7284$ which is too small (because $\frac{5}{6} \approx 0.8333$). So we can eliminate choice H.

110

Let's try choice J next and guess that 100 black marbles must be added to the urn. Then there will be $8 + 100 = 108$ black marbles and a total of $30 + 100 = 130$ marbles. So the probability of drawing a black marble will be $\frac{108}{130} \approx 0.8308$ which is still too small but closer to the correct answer. So we can eliminate choice J and the answer is probably K.

Let's now just confirm that choice K is the answer. If we guess that 102 black marbles must be added to the urn, then there will be $8 + 102 = 110$ black marbles and $30 + 102 = 132$ marbles in total. So the probability of drawing a black marble will be $\frac{110}{132} \approx 0.8333$ which is correct. So the answer is choice **K**.

 * **Algebraic solution:** Let x be the number of black marbles we must add to the urn. Note that x will also be added to the total so that the probability of drawing a black marble will become $\frac{8+x}{30+x}$. We set this equal to $\frac{5}{6}$ and cross multiply to get the following.

$$\frac{8+x}{30+x} = \frac{5}{6}$$

$$6(8 + x) = 5(30 + x)$$

We multiply on each side of this last equation to get $48 + 6x = 150 + 5x$. Subtracting $5x$ from each side yields $48 + x = 150$, and finally, subtracting 48 from each side gives us $x = 150 - 48 = 102$, choice **K**.

123. The average (arithmetic mean) of 4 numbers is x. If one of the numbers is y, what is the average of the remaining 3 numbers in terms of x and y ?

   A. $\frac{x}{4}$

   B. $4x - y$

   C. $\frac{3x-y}{4}$

   D. $\frac{4x-y}{3}$

   E. $\frac{4y-x}{3}$

Solution by picking numbers: Let's let $y = 5$, and we'll let the other 3 numbers be 1, 2, and 4. We then have that $x = \frac{1+2+4+5}{4} = \frac{12}{4} = 3$, and the average of the remaining 3 numbers is $\frac{1+2+4}{3} = \frac{7}{3} \approx \mathbf{2.333}$. We now substitute $x = 3$ and $y = 5$ into all five answer choices.

A. $3 / 4 = .75$
B. $4*3 - 5 = 12 - 5 = 7$
C. $(3*3 - 5) / 4 = (9 - 5) / 4 = 4 / 4 = 1$

D. $(4*3 - 5) / 3 = (12 - 5) / 3 = 7 / 3 \approx 2.333$
E. $(4*5 - 3) / 3 = (20 - 3) / 3 = 17 / 3 \approx 5.667$

Since A, B, C and E are incorrect we can eliminate them. Therefore, the answer is choice **D**.

* Solution by changing averages to sums: We use the formula

$$\text{Sum} = \text{Average} \cdot \text{Number}$$

So the **Sum** of the 4 numbers is $4x$. The sum of the remaining 3 numbers (after removing y) is $4x - y$. We now use the above formula again in the form

$$\text{Average} = \frac{\text{Sum}}{\text{Number}}.$$

So the Average of the remaining 3 numbers is $\frac{4x - y}{3}$, choice **D**.

124. Let a, b and c be numbers with $a < b < c$ such that the average of a and b is 2, the average of b and c is 4, and the average of a and c is 3. What is the average of a, b and c ?

F. 1
G. 2
H. 3
J. 4
K. 5

* We change the averages to sums using the formula

$$\text{Sum} = \text{Average} \cdot \text{Number}$$

$$\text{So } a + b = 4$$
$$b + c = 8$$
$$a + c = 6$$

Adding these equations gives us $2a + 2b + 2c = 18$ so that $a + b + c = 9$. Finally, we divide by 3 to get that the average of a, b and c is $\frac{9}{3} = \mathbf{3}$.

112

LEVEL 4: TRIGONOMETRY

125. Whenever $\frac{\sin x}{\tan x}$ is defined, it is equivalent to:

A. $\cos x$

B. $\sin x$

C. $\frac{1}{\cos x}$

D. $\frac{1}{\sin x}$

E. $\frac{1}{\cos^2 x}$

* Recall that $\tan x = \frac{\sin x}{\cos x}$. So we have

$$\frac{\sin x}{\tan x} = \frac{\sin x}{\left(\frac{\sin x}{\cos x}\right)} = \sin x \div \frac{\sin x}{\cos x} = \sin x \cdot \frac{\cos x}{\sin x} = \cos x$$

This is choice **A**.

Note: For a review of the basic trigonometry used here see the solution to problem 61.

Solution by picking a number: Let's plug a value in for x in our calculator, say $x = 15$. In our calculator we input (sin 15) / (tan 15) and we get an output of approximately **0.9659**.

We now substitute 15 for x into each answer choice.

A. $\cos 15° \approx 0.9659$

B. $\sin 15° \approx 0.2588$

C. $\frac{1}{\cos 15°} \approx 1.035$

D. $\frac{1}{\sin 15°} \approx 3.8637$

E. $\frac{1}{\cos^2 15°} \approx 1.0718$

Since B, C, D and E are incorrect we can eliminate them. Therefore, the answer is choice **A**.

126. If $0 \le x \le 90°$ and $\cos x = \frac{5}{13}$, then $\tan x = ?$

    F. $\frac{5}{13}$

    G. $\frac{5}{12}$

    H. $\frac{13}{12}$

    J. $\frac{12}{5}$

    K. $\frac{13}{5}$

* Let's draw a picture. We begin with a right triangle and label one of the angles x.

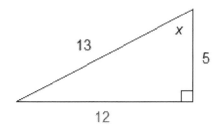

Since $\cos x = \frac{\text{ADJ}}{\text{HYP}}$, we label the leg adjacent to x with a 5 and the hypotenuse with 13. We use the Pythagorean triple 5, 12, 13 to see that the other side is 12.

Finally, $\tan x = \frac{\text{OPP}}{\text{ADJ}} = \frac{12}{5}$, choice **J.**

Notes: (1) The most common Pythagorean triples are 3,4,5 and 5, 12, 13. Two others that may come up are 8, 15, 17 and 7, 24, 25.

(2) If you don't remember the Pythagorean triple 5, 12, 13, you can use the Pythagorean Theorem which says that if a right triangle has legs of length a and b, and a hypotenuse of length c, then $c^2 = a^2 + b^2$.

In this problem we have $5^2 + b^2 = 13^2$. So $25 + b^2 = 169$. Subtracting 25 from each side of this equation gives $b^2 = 169 - 25 = 144$. So $b = 12$.

(3) The equation $b^2 = 144$ would normally have two solutions: $b = 12$ and $b = -12$. But the length of a side of a triangle cannot be negative, so we reject -12.

(4) If necessary, review the basic trigonometry given after the solution to problem 61.

127. In △*ABC* below, *AB* = 12 inches. To the nearest tenth of an inch, *BC* = ?

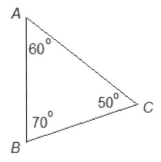

A. 9.8
B. 10.6
C. 13.5
D. 13.6
E. 13.9

* Solution using the Law of Sines: Using the law of sines we have that $\frac{AB}{\sin C} = \frac{BC}{\sin A}$ so that $\frac{12}{\sin 50°} = \frac{BC}{\sin 60°}$. So we have $BC = \frac{12}{\sin 50°} \cdot \sin 60°$. We type this last expression into our calculator to get approximately 13.5661905. which we round to 13.6, choice **D.**

Notes: (1) The **law of sines** says $\frac{a}{\sin A} = \frac{b}{\sin B} = \frac{c}{\sin C}$ where *A*, *B* and *C* are the angles of the triangle, and *a*, *b* and *c* are the lengths of the sides opposite these angles (in that order).

In this problem we used angles *A* and *C*, and their opposite sides.

(2) Make sure that your calculator is in degree mode. Otherwise you will get the incorrect answer 13.9 (choice E).

If you are using a TI-84 (or equivalent) calculator press MODE and on the third line, make sure that DEGREE is highlighted. If it is not, scroll down and select it. If possible, do not alter this setting until you are finished taking your ACT.

Solution using only basic trigonometry: Let's draw altitude *BD* to base *AC* to form two right triangles.

115

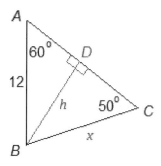

Notice that I labeled side AB with its length 12, and BD and BC with unknown lengths h and x, respectively.

Now using basic trigonometry we have $\sin 60° = \frac{h}{12}$, so that

$$h = 12 \sin 60° \approx 10.3923.$$

We use basic trigonometry once more to get $\sin 50° = \frac{h}{x}$. So we have $x \sin 50° = h$, and therefore $x = \frac{h}{\sin 50°} \approx 13.5661905.$

We round this last result to 13.6, choice **D**.

Note: Triangle ABD is actually a 30, 60, 90 degree triangle. So we could have found h by just taking half of 12 and then multiplying by $\sqrt{3}$ to get $\frac{12\sqrt{3}}{2} = 6\sqrt{3} \approx 10.3923$ (see the end of the solution to problem 62 for details).

128. The vertex of $\angle P$ is the origin of the standard (x, y) coordinate plane. One ray of $\angle P$ is the positive x-axis. The other ray, $\overrightarrow{PQ}$, is positioned so that $\tan A < 0$ and $\sin A > 0$. In which quadrant, if it can be determined, is point Q ?

    F. Quadrant I
    G. Quadrant II
    H. Quadrant II
    J. Quadrant IV
    K. Cannot be determined from the given information

* $\sin A > 0$ in Quadrants I and II and $\tan A < 0$ in Quadrants II and IV. So point Q must be in Quadrant II, choice **G**.

Note: Many students find it helpful to remember the following diagram.

This diagram tells us which trig functions are positive in which quadrants. The **A** stands for "all" so that all trig functions are positive in the first quadrant. Similarly, **S** stands for "sine," **T** stands for "tangent," and **C** stands for "cosine."

So, for example, if an angle A in **standard position** (this just means that its initial side is the positive x-axis) has its terminal side in the second quadrant, then $\sin A > 0$, $\tan A < 0$ and $\cos A < 0$.

Exercise: Let A be an angle in standard position with terminal side in Quadrant I. For each of the six trig functions, determine if applying that function to A will give a positive or negative answer.

Repeat this exercise for Quadrants II, III and IV.

LEVEL 5: NUMBER THEORY

129. What is the sum of the first 5 terms of the arithmetic sequence in which the 7th term is 10 and the 11th term is 16?

   A. 18.5
   B. 20
   C. 22.5
   D. 25
   E. 27.5

* Solution using a linear equation: Recall from the end of the solution to problem 2 that we can find the common difference of this arithmetic sequence with the computation

$$d = \frac{16-10}{11-7} = \frac{6}{4} = 1.5.$$

So the sixth term of the sequence is $10 - 1.5 = 8.5$, the fifth term is $8.5 - 1.5 = 7$, the fourth term is $7 - 1.5 = 5.5$, the third term $5.5 - 1.5 = 4$, the second term $4 - 1.5 = 2.5$, and the first term $2.5 - 1.5 = 1$.

So the sum of the first 5 terms is $1 + 2.5 + 4 + 5.5 + 7 = 20$, choice **B**.

Remarks: (1) Note that we identified the given terms of the sequence with points on a line. The x-coordinates are the term numbers and the y-coordinates are the terms themselves.

In this question the two points are $(7,10)$ and $(11,16)$.

The common difference is then just the slope of the line passing through these two points.

(2) In an arithmetic sequence, we always add (or subtract) the same number to get from one term to the next. This can be done by moving forwards or backwards through the sequence.

Solution using the arithmetic sequence formula: Recall that the nth term of an arithmetic sequence is given by

$$a_n = a_1 + (n - 1)d$$

In this question we are given that $a_7 = 10$ and $a_{11} = 16$. So we get the system of equations

$$16 = a_1 + 10d$$
$$10 = a_1 + 6d$$

We subtract these equations to get $6 = 4d$. Dividing each side of this last equation by 6 gives $d = \frac{6}{4} = 1.5$. We can now substitute this in to either of the two original equations to find the first term of the sequence. Using the first equation we have

$$16 = a_1 + 10(1.5) = a_1 + 15.$$

So $a_1 = 16 - 15 = 1$. Now using the common difference $d = 1.5$ we can easily get the first 5 terms of the sequence.

$$1, 2.5, 4, 5.5, 7$$

We add these up to get $1 + 2.5 + 4 + 5.5 + 7 = 20$, choice **B**.

130. In the equation $\log_4 2 + \log_4 8 = \log_6 x^2$, what is the positive real value of x ?

    F. 6
    G. 7
    H. 8
    J. 9
    K. 10

* We have $\log_4 2 + \log_4 8 = \log_4 16 = 2$ using a basic law of logarithms (or our calculator). So $2 = \log_6 x^2$. Therefore $x^2 = 6^2$ and so $x = 6$, choice **F**.

Notes: (1) See the end of the solution to problem 97 for some basic information about logarithms.

(2) The sum of two logarithms with the same base can be written as a product under a single logarithm. For example, since $(2)(8) = 16$, we have that $\log_4 2 + \log_4 8 = \log_4 16$.

(3) $\log_4 16 = 2$ because $16 = 4^2$. We simply changed from logarithmic form to exponential form here. For the same reason, if $2 = \log_6 x^2$, then $x^2 = 6^2$.

(4) The equation $x^2 = 6^2$ has the two solutions $x = 6$ and $x = -6$. Since the question is asking for the positive real value of x, we only take $x = 6$.

Laws of Logarithms: Here is a brief review of the basic laws of logarithms.

Law	Example
$\log_b 1 = 0$	$\log_2 1 = 0$
$\log_b b = 1$	$\log_6 6 = 1$
$\log_b x + \log_b y = \log_b(xy)$	$\log_5 7 + \log_5 2 = \log_5 14$
$\log_b x - \log_b y = \log_b(\frac{x}{y})$	$\log_3 21 - \log_3 7 = \log_3 3 = 1$
$\log_b x^n = n\log_b x$	$\log_8 3^5 = 5\log_8 3$

131. For every positive 3-digit number, z, with hundreds digit a, tens digit b, and units digit c, let w be the 3-digit number formed by interchanging a and c (leaving b fixed). Which of the following expressions is equivalent to $w - z$?

    A. $99(c - a)$
    B. $99(a - c)$
    C. $99c - a$
    D. $99a - c$
    E. 0

Solution by picking numbers: Let's let $a = 2$, $b = 3$ and $c = 4$. It follows that $z = 234$, $w = 432$ and $w - z = 432 - 234 = \mathbf{198}$. We now substitute our values for a and c into each answer choice.

    A. $99(4 - 2) = 198$
    B. $99(2 - 4) = -198$
    C. $99(4) - 2 = 394$
    D. $99(2) - 4 = 194$
    E. 0

Since B, C, D and E are incorrect we can eliminate them. Therefore, the answer is choice **A.**

* Algebraic solution: We have

$$z = 100a + 10b + c \text{ and } w = 100c + 10b + a.$$
$$w - z = (100c + 10b + a) - (100a + 10b + c) =$$
$$100c + 10b + a - 100a - 10b - c = 99c - 99a = 99(c - a).$$

This is choice **A.**

Remark: Observe how we used the expanded form of a number here. For example, the number abc in standard form can be written in expanded form as $100a + 10b + c$.

It's usually helpful to put a whole number into expanded form when we want to manipulate the digits individually.

132. The first and second terms of a geometric sequence are k and bk, in that order. What is the 500th term of the sequence?

    F. $(bk)^{500}$
    G. $(bk)^{499}$
    H. $b^{501}k$
    J. $b^{500}k$
    K. $b^{499}k$

Solution by listing: We can find the **common ratio** of a geometric sequence by dividing any term by the previous term. In this problem the common ratio is $\frac{bk}{k} = b$. So to get from one term to the next we multiply by b. The first term is k, the second term is bk, the third term is $bbk = b^2k$, the fourth term is b^3k, and so on. So it appears that the 500th term is $b^{499}k$, choice **K.**

Remark: (1) In a geometric sequence, you always multiply (or divide) by the same number to get from one term to the next. This can be done by moving forwards or backwards through the sequence.

(2) Notice that we simply observed that the exponent is always one less than the term number. For example, the **4**th term is b^3k.

Advanced material: A **geometric sequence** is a sequence of numbers such that the quotient r between consecutive terms is constant. The number r is called the **common ratio** of the geometric sequence.

Here is an example of a geometric sequence: $20, 10, 5, \frac{5}{2}, \frac{5}{4}, \ldots$
In this example the common ratio is $r = \frac{10}{20} = \frac{1}{2}$.

Geometric sequence formula: $g_n = g_1 \cdot r^{n-1}$

In the above formula, g_n is the nth term of the sequence. For example, g_1 is the first term of the sequence.

Note: In the geometric sequence $20, 10, 5, \frac{5}{2}, \frac{5}{4}, \dots$ we have that $g_1 = 20$ and $r = \frac{1}{2}$. Therefore

$$g_n = 20 \left(\frac{1}{2}\right)^{n-1} .$$

This expression cannot be simplified!

* Quick solution using the geometric sequence formula: The first term of the sequence is $g_1 = k$ and the common ratio is $r = \frac{bk}{k} = b$. So we have $g_n = kb^{n-1}$. So $g_{500} = kb^{499} = b^{499}k$, choice **K**.

133. The quantity $\sqrt[n]{5^x}$ is defined when n is an integer greater than 2 and x is any nonzero real number. Which of the following is a relationship between n and x that will always make $\sqrt[n]{5^x}$ a positive integer?

    A. x is less than n

    B. n is less than x

    C. $n + x = 1$

    D. $\frac{n}{x}$ is a positive integer

    E. $\frac{x}{n}$ is a positive integer

Solution by picking numbers: Let's take a look at choice C and choose values for x and n so that $n + x = 1$, let's say $n = 3$ and $x = -2$. Then we have $\sqrt[n]{5^x} = \sqrt[3]{5^{-2}} \approx .342$ which is not an integer. Since -2 is less than 3 and $3 + (-2) = 1$, we can eliminate choices A and C.

Let's now focus on choice B and choose values for x and n so that n is less than x, let's say $n = 3$ and $x = 4$. Then $\sqrt[n]{5^x} = \sqrt[3]{5^4} \approx 8.55$ which is again not an integer. So we can eliminate choice B.

Looking at choice D, let's choose values for x and n so that $\frac{n}{x}$ is a positive integer, say $n = 4$ and $x = 2$. Then $\sqrt[n]{5^x} = \sqrt[4]{5^2} \approx 2.24$, not an integer. So we can eliminate choice D, and the answer must be choice **E**.

Note: You can find a general root function in your TI-84 (or equivalent) calculator by pressing the MATH button. So for example, to compute

121

$\sqrt[4]{5^2}$, type 4, press the MATH button, press 5 to select the general root function, type 5^2 (or 5 followed by the x^2 button) and press ENTER.

* **Quick solution:** $\sqrt[n]{5^x} = 5^{\frac{x}{n}}$ will be a positive integer precisely when $\frac{x}{n}$ is a positive integer, choice **E**.

More Laws of Exponents: Review of negative and fractional exponents.

Law	Example
$x^{-1} = \dfrac{1}{x}$	$3^{-1} = \dfrac{1}{3}$
$x^{-a} = \dfrac{1}{x^a}$	$9^{-2} = \dfrac{1}{81}$
$x^{\frac{1}{n}} = \sqrt[n]{x}$	$x^{\frac{1}{3}} = \sqrt[3]{x}$
$x^{\frac{m}{n}} = \sqrt[n]{x^m} = \left(\sqrt[n]{x}\right)^m$	$x^{\frac{9}{2}} = \sqrt{x^9} = \left(\sqrt{x}\right)^9$

134. For all positive integers k, which of the following is a correct ordering of the terms k^{k+1}, $k^{1+k!}$, and $(k+1)^{(k+1)!}$?

    F. $k^{k+1} \geq k^{1+k!} \geq (k+1)^{(k+1)!}$
    G. $k^{1+k!} \geq k^{k+1} \geq (k+1)^{(k+1)!}$
    H. $(k+1)^{(k+1)!} \geq k^{1+k!} \geq k^{k+1}$
    J. $k^{k+1} \geq (k+1)^{(k+1)!} \geq k^{1+k!}$
    K. $k^{1+k!} \geq (k+1)^{(k+1)!} \geq k^{k+1}$

* **Solution by picking a number:** Let's choose a value for k, say $k = 3$. Then $k^{k+1} = 3^4 = 81$, $k^{1+k!} = 3^7 = 2187$, and $(k+1)^{(k+1)!} = 4^{24}$ (a tremendous number). So we see that $(k+1)^{(k+1)!} \geq k^{1+k!} \geq k^{k+1}$, choice **H**.

Notes: (1) You can do all of these computations in your calculator.

(2) $k! = 1 \cdot 2 \cdot 3 \cdots k$.

So for example, $4! = 1 \cdot 2 \cdot 3 \cdot 4 = 24$.

Algebraic solution: $(k+1)! = (k+1)k! = k \cdot k! + k! \geq 1 + k!$ and $k + 1 \geq k$. So $(k+1)^{(k+1)!} \geq k^{1+k!}$.

Also $1 + k! \geq 1 + k = k + 1$. So $k^{1+k!} \geq k^{k+1}$.

Therefore, the answer is choice **H**.

135. The sum of an infinite geometric series with first term a and common ratio r with $-1 < r < 1$ is given by $\frac{a}{1-r}$. The sum of a given infinite geometric series is 160 and the common ratio is $\frac{1}{4}$. What is the third term of this series?

    A. 7.5
    B. 12
    C. 17.5
    D. 22
    E. 27.5

* We are given that $r = \frac{1}{4}$, so that $\frac{a}{1-r} = \frac{a}{1-.25} = \frac{a}{.75}$. We are also given that $\frac{a}{1-r} = 160$ so that $\frac{a}{.75} = 160$. Multiplying each side of this equation by .75 gives $a = 160(.75) = 120$.

So the first term of the series is 120. To get each term after that we multiply by the common ratio. So the second term is $\frac{1}{4}(120) = 30$ and the third term is $\frac{1}{4}(30) = 7.5$, choice **A**.

Notes: (1) A geometric series is the sum of the terms of a geometric sequence.

(2) See problem 132 for more information on geometric sequences.

136. If n is a positive integer and $k = n^3 - n$, which of the following statements about k must be true for all values of n?

    I. k is a multiple of 3
    II. k is a multiple of 4
    III. k is a multiple of 6

    F. I only
    G. II only
    H. III only
    J. I and III only
    K. I, II, and III

Solution by picking numbers: Let's try some values for n.

$n = 2$. Then $k = 6$. This shows that II can be false. So we can eliminate choices G and K.

$n = 3$. Then $k = 24$. This is divisible by both 3 and 6.

$n = 4$. Then $k = 60$. This is again divisible by both 3 and 6.

The evidence seems to suggest that the answer is choice **J**.

Remark: This method is a bit risky. Since this is a Level 5 problem, there is a chance that some large value of n might provide a counterexample. In this case, it turns out not to be the case, and the answer is in fact choice J.

* Advanced solution:

$$n^3 - n = n(n^2 - 1) = n(n - 1)(n + 1) = (n - 1)n(n + 1).$$

Thus $n^3 - n$ is a product of 3 consecutive integers. In a product of 2 consecutive integers, one of the integers must be divisible by 2 (even). In a product of 3 consecutive integers, one of the integers must be divisible by 3. Therefore, $n^3 - n$ is divisible by both 2 and 3, and thus it is divisible by 6. As in the first solution above, $n = 2$ shows that the expression does **not** have to be divisible by 4. Thus, the answer is choice **J**.

Exercise for the very advanced student: Let k be any positive integer. Show that the product of k consecutive integers is divisible by k!

LEVEL 5: ALGEBRA AND FUNCTIONS

137. In the equation $x^2 + bx + c = 0$, b and c are integers. The <u>only</u> possible value for x is –5. What is the value of $b + c$?

   A.   15
   B.   –15
   C.   25
   D.   –25
   E.   35

* Quick solution: We are given that both solutions of the given quadratic equation are –5. So the sum of the solutions is $-b = -5 + (-5) = -10$ and the product of the solutions is $c = (-5)(-5) = 25$. So $b + c = 10 + 25 = 35$, choice **E**.

Notes: (1) If r and s are the solutions of the quadratic equation $x^2 + bx + c = 0$, then $b = -(r + s)$ and $c = rs$.

Alternate solution: Since –5 is the only solution of the equation, we have that $(x + 5)$ is the only factor. So

$$x^2 + bx + c = (x + 5)(x + 5) = x^2 + 10x + 25.$$

So $b = 10$, $c = 25$, and $b + c = 10 + 25 = 35$, choice **E**.

138. Given $f(x) = \frac{2x-3}{x+5}$ and $g(x) = x^2 + 1$, which of the following is an expression for $(f \circ g)(x)$?

    F. $\frac{2x^2-1}{x^2+6}$

    G. $\frac{2x^2-1}{x+5}$

    H. $\frac{2x-3}{x^2+6}$

    J. $\frac{2x^2-2}{x^2+6}$

    K. $\left(\frac{2x-3}{x+5}\right)^2 + 1$

* $(f \circ g)(x) = f(g(x)) = f(x^2 + 1) = \frac{2(x^2+1)-3}{(x^2+1)+5} = \frac{2x^2+2-3}{x^2+6} = \frac{2x^2-1}{x^2+6}.$

This is choice **F**.

Notes: (1) $(f \circ g)$ is called the **composition** of the functions f and g. We obtain $(f \circ g)(x)$ by substituting the whole function g in place of x in the function f.

So in this problem, we replace each of the two x's in the function f by the expression $x^2 + 1$.

(2) This problem can also be solved by picking a number for x, but this would be quite time consuming.

139. A parabola with axis of symmetry $x = 2$ crosses the x-axis at $(2 + \sqrt{3}, 0)$. At what other point, if any, does the parabola cross the x-axis?

    A. $\left(-2 - \sqrt{3}, 0\right)$
    B. $\left(-2 + \sqrt{3}, 0\right)$
    C. $\left(2 - \sqrt{3}, 0\right)$
    D. No other point
    E. Cannot be determined from the given information

* A parabola is symmetrical about its axis of symmetry. So if there is an x-intercept to the right of the axis, there will be another x-intercept to the left of the axis the same distance away. Therefore $\left(2 - \sqrt{3}, 0\right)$ is an x-intercept of the parabola, choice **C**.

Note: Let's look at a picture of the parabola with the axis of symmetry and the two x-intercepts.

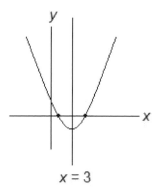

$x = 3$

Note that the vertical line labeled "$x = 3$" is the axis of symmetry. To the right of the axis of symmetry on the x-axis is the point $(2 + \sqrt{3}, 0)$, and to the left of the axis of symmetry (but to the right of the y-axis) is the point $(2 - \sqrt{3}, 0)$.

140. What is the smallest possible value for the product of 2 real numbers that differ by 7 ?

   F. −49
   G. −12.25
   H. − 7.25
   J.   0
   K.   12.25

* Algebraic solution: If we let x be the smaller number, then since the numbers differ by 7, the larger number is $x + 7$. The product of the two numbers is then $y = x(x + 7) = x^2 + 7x$.

The expression will be smallest when $x = -\dfrac{7}{2(1)} = -3.5$ (see Remark (2) below). So the smallest possible product is $y = (-3.5)^2 + 7(-3.5) = -12.25$, choice **G.**

Remarks: (1) Note that $(x + 7) - x = 7$. So x and $x + 7$ differ by 7.

(2) The general form for a quadratic function is

$$y = ax^2 + bx + c.$$

The graph of this function is a parabola whose vertex has x-coordinate

$$-\frac{b}{2a}$$

126

The parabola opens upwards if $a > 0$ and downwards if $a < 0$.

In the equation $y = x^2 + 7x$, we have $a = 1$, $b = 7$ and $c = 0$. So we have

$$-\frac{b}{2a} = -\frac{7}{2(1)} = -3.5.$$

Note that the parabola opens upwards so that we get a minimum at $x = -3.5$.

141. The equation $3^{x^2-2x+8} = 243$ has two solutions. Let a be the sum of these solutions and let b be the product of these solutions. What is $a - b$?

    A. -2
    B. -1
    C. 0
    D. 1
    E. 2

* $243 = 3^5$. So we have $3^{x^2-2x+8} = 3^5$. Therefore, $x^2 - 2x + 8 = 5$. Subtracting 5 from each side of this equation gives $x^2 - 2x + 3 = 0$. The sum of the solutions is the negative of the coefficient of the x term so that $a = 2$. The product of the solutions is the constant term so that $b = 3$.

So $a - b = 2 - 3 = -1$, choice **B**.

Notes: (1) If r and s are the solutions of the quadratic equation $x^2 + bx + c = 0$, then $b = -(r + s)$ and $c = rs$. So in this problem the sum of the 2 solutions is 7 and the product of the 2 solutions is 3.

(2) Yes, you can also solve the equation $x^2 - 2x + 3 = 0$ by completing the square or using the quadratic formula, but this is very time consuming. It is much better to use the solution given above.

(3) When both sides of an equation have the same base, the exponents on each side of the equation are equal as well. In this problem each side of the original equation has a base of 3. So we get $x^2 - 2x + 8 = 5$.

142. Consider the equation $\sqrt{y} - \sqrt{-x} = 5\sqrt{-x}$, where x is a negative real number and y is a positive real number. What is y in terms of x ?

    F. $-36x$
    G. $-12x$
    H. $- 6x$
    J. $6x$
    K. $12x$

* We add $\sqrt{-x}$ to each side of the given equation to get $\sqrt{y} = 6\sqrt{-x}$. We now square each side of this equation to get $y = 36(-x) = -36x$, choice **F**.

Note: This problem can also be solved by picking a value for x. A good initial guess for x would be -4 (do you see why?). I leave the details to the reader.

143. The solution set of which of the following equations is the set of real numbers that are 7 units from -2 ?

    A. $|x + 2| = 7$
    B. $|x - 2| = 7$
    C. $|x + 2| = -7$
    D. $|x - 7| = 2$
    E. $|x + 7| = 2$

* We are looking for all real numbers x such that the distance between x and -2 is 7. The distance between x and -2 can be written

$$|x - (-2\ )| = |x + 2|.$$

So the answer is $|x + 2| = 7$, choice **A**.

An advanced lesson on absolute value and distance:

Geometrically, $|x - y|$ is the distance between x and y. In particular, $|x - y| = |y - x|$.

Examples: $|5 - 3| = |3 - 5| = 2$ because the distance between 3 and 5 is 2.

If $|x - 3| = 7$, then the distance between x and 3 is 7. So there are two possible values for x. They are $3 + 7 = 10$, and $3 - 7 = -4$. See the figure below for clarification.

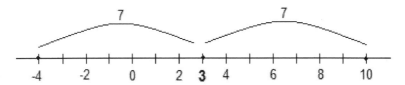

If $|x - 3| < 7$, then the distance between x and 3 is less than 7. If you look at the above figure you should be able to see that this is all x satisfying $-4 < x < 10$.

If $|x - 3| > 7$, then the distance between x and 3 is greater than 7. If you look at the above figure you should be able to see that this is all x satisfying $x < -4$ or $x > 10$

Algebraically, we have the following. For $c > 0$,

$$|x| = c \text{ is equivalent to } x = c \text{ or } x = -c$$

$$|x| < c \text{ is equivalent to } -c < x < c$$

$$|x| > c \text{ is equivalent to } x < -c \text{ or } x > c.$$

Let's look at the same examples as before algebraically.

Examples: If $|x - 3| = 7$, then $x - 3 = 7$ or $x - 3 = -7$. So $x = 10$ or $x = -4$.

If $|x - 3| < 7$, then $-7 < x - 3 < 7$. So $-4 < x < 10$.

If $|x - 3| > 7$, then $x - 3 < -7$ or $x - 3 > 7$. So $x < -4$ or $x > 10$.

144. The operation $\ominus$ is defined by the following:
$$x \ominus y = 7 + x - y + 2xy$$
If $x \ominus y = y \ominus x$, then which of the following describes all the possible values of x and y ?

    F. They have opposite signs.
    G. They are equal.
    H. They are both positive.
    J. They are both negative.
    K. They can have any values.

* We are given that $x \ominus y = 7 + x - y + 2xy$ and by interchanging x and y we have $y \ominus x = 7 + y - x + 2yx$. So we are looking for all solutions of
$$7 + x - y + 2xy = 7 + y - x + 2yx.$$
But this equation is always true! So the answer is choice **K**.

Remark: Here we have used the fact that the set of real numbers is **commutative** for addition and multiplication. For more information about commutativity see the end of the solution to problem 43.

LEVEL 5: GEOMETRY

145. For the triangles in the figure below, which of the following ratios of side lengths is equivalent to the ratio of the perimeter of ΔCBA to the perimeter of ΔMAC ?

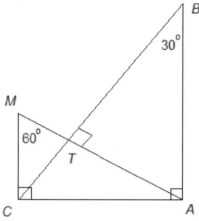

- **A.** $AB:CA$
- **B.** $AB:AM$
- **C.** $AB:BC$
- **D.** $AB:CM$
- **E.** $BC:CM$

* We redraw the two right triangles next to each other so that congruent angles match up.

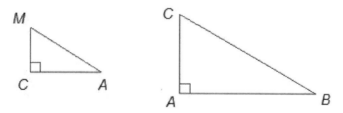

Now we simply observe that side AB on the larger triangle corresponds to side CA on the smaller angle. So the answer is choice **A.**

Remarks: (1) Both triangles have the same angle measures. The measures of angles CMA and ACB are $60°$ and the measures of angles MAC and CBA are $30°$. Therefore, the two triangles are **similar.**

(2) Corresponding sides of similar triangles are in proportion. So for example $AB:CA$ is equal to $CA:MC$, or equivalently $\frac{AB}{CA} = \frac{CA}{MC}$.

(3) To see why $\frac{AB}{CA} = \frac{\text{perimeter of } \Delta CBA}{\text{perimeter of } \Delta MAC}$, first observe that

$$\frac{AB}{CA} = \frac{CA}{MC} \text{ and } \frac{AB}{CA} = \frac{BC}{AM}, \text{ so that } CA = \frac{(AB)(MC)}{CA} \text{ and } BC = \frac{(AB)(AM)}{CA}.$$

And so we have $\frac{AB+BC+CA}{CA+AM+MC} = \frac{AB+\frac{(AB)(AM)}{CA}+\frac{(AB)(MC)}{CA}}{CA+AM+MC} = \frac{AB(CA+AM+MC)}{CA(CA+AM+MC)} = \frac{AB}{CA}.$

Note that the second expression in the above equation is a complex fraction. To simplify this complex fraction, we multiplied the numerator and denominator by CA.

146. An isosceles right triangle, T_1, has a hypotenuse of length $10\sqrt{2}$ units. The vertices of a second right triangle, T_2, are the midpoints of the sides of T_1. The vertices of a third right triangle, T_3, are the midpoints of the sides of T_2. This process continues indefinitely, with the vertices of T_{k+1} being the midpoints of the sides of T_k for each integer $k > 0$. What is the sum of the areas, in square units, of $T_1, T_2, ...$?

F. $\frac{25}{3}$

G. $\frac{50}{3}$

H. $\frac{100}{3}$

J. $\frac{200}{3}$

K. 200

* Let's draw a picture of the first few triangles.

131

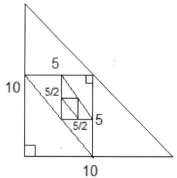

Note that an isosceles right triangle is the same as a 45, 45, 90 triangle. Since the hypotenuse of T_1 has length $10\sqrt{2}$, each leg has length 10, and therefore the area of T_1 is $\frac{1}{2}(10)(10) = 50$.

The legs of T_2 have lengths that are half the lengths of the legs of T_1. So each leg of T_2 has length 5, and the area of T_2 is $\frac{1}{2}(5)(5) = \frac{25}{2}$.

Let's do one more in case you don't see the pattern yet. The legs of T_3 have length $\frac{5}{2}$ so that the area of T_3 is $\frac{1}{2}(\frac{5}{2})(\frac{5}{2}) = \frac{25}{8}$.

So we want to compute the sum $50 + \frac{25}{2} + \frac{25}{8} + \cdots$

This is a geometric series with first term $g_1 = 50$ and common ratio $r = \frac{25}{2} \div 50 = \frac{1}{4}$. So the sum is $\frac{g_1}{1-r} = \frac{50}{1-\frac{1}{4}} = \frac{50}{\frac{3}{4}} = 50 \cdot \frac{4}{3} = \frac{200}{3}$, choice **J.**

Note: It is okay to use a calculator for all of these computations.

147. A circular disk is cut out of a larger circular disk, as shown below, so that the area of the piece that remains is the same as the area of the cutout. If the radius of the larger circle is R, what is the circumference of the cutout, in terms of R ?

A. R
B. $R\pi$
C. $R\sqrt{2}$
D. $R\pi\sqrt{2}$
E. $2R\pi\sqrt{2}$

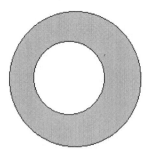

132

* Let's let r be the radius of the cutout. Then the area of the small circle is πr^2 and the area of the piece that remains (shaded part) is $\pi R^2 - \pi r^2$. We are given that these two areas are equal so that $\pi R^2 - \pi r^2 = \pi r^2$. Adding πr^2 to each side of this equation gives $\pi R^2 = 2\pi r^2$. We now divide each side of this equation by 2π to get $\frac{R^2}{2} = r^2$. Taking the positive square root of each side yields $\frac{R}{\sqrt{2}} = r$. The circumference of the cutout is therefore $2\pi r = 2\pi \frac{R}{\sqrt{2}}$ This is equal to $R\pi\sqrt{2}$ (see Notes (5), (6) below), choice **D.**

Notes: (1) For basic information about circles see the information at the end of the solution to problem 86.

(2) The **circumference** of a circle with radius r is $C = 2\pi r$ and the **area** of a circle with radius r is $A = \pi r^2$.

(3) In this problem we are looking at two circles. The large circle is referred to as the "larger circular disk" in the problem and it has radius R. Therefore, its area is πR^2.

The small circle is referred to as both "a circular disk" and "the cutout" in the problem. In the solution we named its radius r. Therefore, its area is πr^2 and its circumference is $2\pi r$.

(4) The shaded region is referred to as "the piece that remains" in the problem and we can find its area by subtracting the area of the large circle minus the area of the small circle. This is a typical computation in **problems with shaded regions**. In this case the area of the shaded region is $\pi R^2 - \pi r^2$.

(5) Once we get the answer $2\pi\frac{R}{\sqrt{2}}$, we see that this is not an answer choice. We can simply ignore the R (since it's in every answer choice) and put $2\pi\frac{1}{\sqrt{2}}$ into our calculator. Then do the same with all the answer choices until we get the one that "matches up." This will be choice D.

(6) If you really want to make the answer $2\pi\frac{R}{\sqrt{2}}$ "match up" with one of the answer choices by hand, you can **rationalize the denominator** by multiplying both the numerator and denominator by $\sqrt{2}$. The computation looks like this: $2\pi\frac{R}{\sqrt{2}} \cdot \frac{\sqrt{2}}{\sqrt{2}} = 2\pi\frac{R\sqrt{2}}{2} = \pi R\sqrt{2} = R\pi\sqrt{2}$.

148. In the figure below, line k has the equation $y = -x$. Line m is below line k, as shown, and m is parallel to k. Which of the following is an equation for line m ?

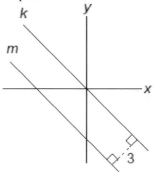

F. $y = -x + 3$
G. $y = -x + \sqrt{3}$
H. $y = -x - 3\sqrt{2}$
J. $y = -3x$
K. $y = -3x + 3$

* Let's add some information to the picture.

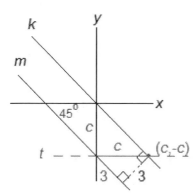

Note that the right triangle in the third quadrant has a 45° angle. We also have the two parallel lines k and m cut by the transversal t. It follows that the smaller right triangle in the fourth quadrant is a 45, 45, 90 triangle (when two parallel lines are cut by a transversal, the alternate interior angles formed have the same measure). So it is isosceles and both legs have length 3. Therefore the hypotenuse of this triangle has length $3\sqrt{2}$ (this side is labeled with a c in the picture).

134

The larger triangle in the fourth quadrant is also isosceles. One way we can see this is by noting that if the x-coordinate of a point on line k is c, then the y-coordinate is $-c$. To plot the point $(c, -c)$ we move right c, then down c (or equivalently, down c, then right c).

It follows that the y-intercept of line m has y-coordinate $-3\sqrt{2}$.

Additionally, since line m is parallel to line k, the slope of line m is -1.

So the equation of line m in $y = mx + b$ form is $y = -x - 3\sqrt{2}$, choice **H**.

Notes: (1) See the end of the solution to problem 128 if you forgot where each quadrant is in the plane.

(2) See problem 88 if you need practice with parallel lines cut by a transversal.

(3) There are two ways to find c. We can simply use the 45, 45, 90 triangle (see the end of problem 62), or we can use the Pythagorean Theorem to get $c^2 = 3^2 + 3^2 = 9 + 9 = 18$. So

$$c = \sqrt{18} = \sqrt{9 \cdot 2} = \sqrt{9}\sqrt{2} = 3\sqrt{2}.$$

(4) See problem 52 for more about the slope-intercept form for the equation of a line.

(5) Parallel lines have the same slope. Since the line $y = -x$ has a slope of -1, the slope of line m is -1 as well.

149. The side lengths of ΔPQR, shown in the figure below, are in inches. One of the 5 points, V, W, X, Y, or Z is the center of a circle that goes through points P, Q, and R. Which point is the center?

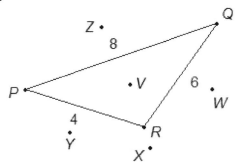

- **A.** V
- **B.** W
- **C.** X
- **D.** Y
- **E.** Z

135

* Solution by process of elimination: If O is the center of a circle that passes through P, Q and R, then $OP = OQ = OR$ since they are all equal to the radius of the circle.

We can eliminate choices A, C and D because R is much closer to V, X and Y than Q is.

We can eliminate choice B because Q is much closer to W than P is.

By process of elimination the answer is choice **E**.

Remark: Observe that the distances from Z to P, Q and R look like they are the same.

150. The circles below, centered at O and O' intersect at A and A', and points O, P, P', and O' are collinear. $OA = 10$, $O'A = 13$, and $PP' = 4$. What is the length of OO'?

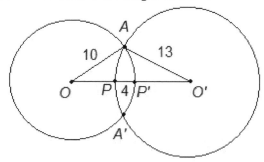

    F. 19
    G. 20
    H. 21
    J. 22
    K. 23

* Notice that OP' is a radius of circle O. So $OP' = OA = 10$. Similarly, $O'P$ is a radius of circle O'. So $O'P = O'A = 13$. So we have

$$OO' = OP' + O'P - PP' = 10 + 13 - 4 = 19.$$

This is choice **F**.

Notes: (1) To see that OP' is a radius of circle O note that P' is a point on the circle. Similarly, P is a point on circle O'.

(2) Note that when we add OP' and $O'P$ we add in PP' twice. So we need to subtract it off once.

151. A square is inscribed in a circle of diameter d. What is the perpendicular distance from the center of the circle to a side of the square, in terms of d ?

A. $\dfrac{d}{2}$

B. $\dfrac{d\sqrt{2}}{4}$

C. $\dfrac{d\sqrt{2}}{2}$

D. d

E. $d\sqrt{2}$

* Let's draw a picture.

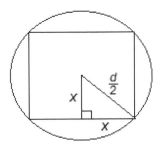

In this picture we have drawn a square inscribed in a circle together with a line segment from the center of the circle to a side of the square, and a radius of the circle from the center of the circle to a vertex of the square.

Note that since the diameter of the circle is d, the radius of the circle is $\dfrac{d}{2}$. Also note that the distance from the center of the circle to a side of the square is half the length of a side of the square. This justifies labeling both this distance and half a side of the square by x in the above figure.

Now there are two ways we can find x:

Method 1: Since both legs of the triangle have length x, an isosceles right triangle is formed. This is the same as a 45, 45, 90 triangle. So $\dfrac{d}{2} = x\sqrt{2}$ and so $x = \dfrac{d}{2\sqrt{2}}$. We can now either use our calculator or rationalize the denominator here to see that $x = \dfrac{d\sqrt{2}}{4}$ (see Notes (5) and (6) at the end of the solution to problem 147). This is choice **B**.

Method 2: We use the Pythagorean Theorem to get

$$x^2 + x^2 = \left(\frac{d}{2}\right)^2$$
$$2x^2 = \frac{d^2}{4}$$
$$x^2 = \frac{d^2}{8}$$
$$x = \frac{d}{2\sqrt{2}}$$

We can now either use our calculator or rationalize the denominator here to see that $x = \frac{d\sqrt{2}}{4}$ (see Notes (5) and (6) at the end of the solution to problem 147). This is choice **B**.

152. In the figure below, $\overline{QS}$ is the shorter diagonal of rhombus $PQRS$ and T is on $\overrightarrow{PS}$. The measure of angle PQS is $x°$. What is the measure of RST, in terms of x ?

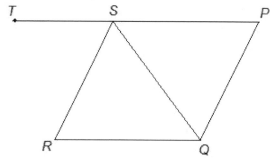

    F. $x°$

    G. $2x°$

    H. $\frac{1}{2}x°$

    J. $90° - x°$

    K. $180° - 2x°$

* The diagonals of a rhombus are perpendicular. So we get the following picture.

138

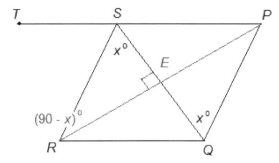

Note that *RS* and *PQ* are parallel lines cut by the transversal *SQ* and ∠*RSQ* and ∠*PQS* are **alternate interior angles**. It follows that the measure of angle *RSQ* is also $x°$.

Since triangle *SER* is a right triangle, it follows that the measure of angle *SRP* is $(90-x)°$.

Since *PQRS* is a rhombus, $SR = PS$ so that ∠*SPR* has the same measure as ∠*SRP*. So the measure of ∠*SPR* is $(90-x)°$.

Now ∠*RST* is an exterior angle to triangle *PRS*. Since **the measure of an exterior angle of a triangle is the sum of the measures of the two opposite interior angles of the triangle**, we have

$$m∠RST = 90 - x + 90 - x = (180 - 2x)°$$

This is choice **K**.

LEVEL 5: PROBABILITY AND STATISTICS

153. One hundred cards numbered 200 through 299 are placed into a bag. After shaking the bag, 1 card is randomly selected from the bag. Without replacing the first card, a second card is drawn. If the first card drawn is 265, what is the probability that both cards drawn have the same tens digit?

A. $\dfrac{1}{8}$

B. $\dfrac{1}{9}$

C. $\dfrac{1}{10}$

D. $\dfrac{1}{11}$

E. $\dfrac{1}{99}$

* After selecting the first card there are now a total of 99 cards left to choose from. There are 9 more cards with a tens digit of 6. So the probability that both cards have the same tens digit is $\frac{9}{99} = \frac{1}{11}$, choice **D**.

Remarks: (1) To compute a simple probability where all outcomes are equally likely, divide the number of "successes" by the total number of outcomes.

In this problem, the total is 99 cards, and 9 of them are "successes."

(2) We started with 100 cards and 1 was already selected. Since we are not replacing the card, the total is $100 - 1 = 99$.

154. A five-digit number is to be formed using each of the digits 1, 2, 3, 4 and 5 exactly once. How many such numbers are there in which the digits 1 and 2 are not next to each other?

    F.   22
    G.   36
    H.   56
    J.   72
    K. 144

Method 1: Start by thinking about where 1 can go. There are 2 cases:

1st Case: 1 is placed at an end. In this case, there are now 3 places where the 2 can go. After the 2 is placed, there are 3 places for the 3, then 2 places for the 4, and then 1 place for the 5. By the counting principle there are $(2)(3)(3)(2)(1) = $ **36** ways to form the five digit number when the 1 is placed at either end (note that the first 2 comes from the fact that we have 2 choices for 1 – the far left or the far right).

2nd Case: 1 is not placed at an end. In this case, there are now 2 places where the 2 can go, and then the rest is the same as case 1. So again by the counting principle there are $(3)(2)(3)(2)(1) = $ **36** ways to form the five digit number when the 1 is **not** placed at either end (note that the first 3 comes from the fact that we have 3 choices for 1 – each of the 3 middle positions).

So adding up the possibilities from cases 1 and 2, we get $36 + 36 = 72$ possibilities all together, choice **J**.

* **Method 2:** Let's first compute the number of ways to place the 1 and 2 with 1 to the left of 2. If the 1 is placed in the leftmost position, then there are 3 places to put the 2 to the right of the 1. If the 1 is placed in the next position to the right, then there are 2 places to put the 2 to the right of the 1. If the 1 is placed in the middle position, then there is 1 place to put the 2 to the right of the 1. Thus, there are $3 + 2 + 1 = 6$ places to put the 1 and 2 with $1 < 2$. By symmetry, there are 6 places to put the 1 and the 2 with $2 < 1$. So all together there are 12 places to put the 1 and 2. Once the 1 and 2 are placed, there are 3 places to put the 3, then 2 places to put the 4, and 1 place to put the 5. By the counting principle the answer is $(12)(3)(2)(1) = 72$, choice **J**.

155. If $t = j + k + m + n + p + q + r$, what is the average (arithmetic mean) of j, k, m, n, p, q, r and t in terms of t ?

    A. $\dfrac{t}{2}$

    B. $\dfrac{t}{3}$

    C. $\dfrac{t}{4}$

    D. $\dfrac{t}{5}$

    E. $\dfrac{t}{6}$

* The average of j, k, m, n, p, q, r and t is

$$\frac{j+k+m+n+p+q+r+t}{8} = \frac{j+k+m+n+p+q+r+j+k+m+n+p+q+r}{8}$$
$$= \frac{2j+2k+2m+2n+2p+2q+2r}{8} = \frac{2(j+k+m+n+p+q+r)}{8} = \frac{2t}{8} = \frac{t}{4}.$$

This is choice **C**.

Alternate solution by picking numbers: Let's let $j = 1$, $k = 2$, $m = 3$, $n = 4$, $p = 5$, $q = 6$ and $r = 7$. Then $t = 28$, and the average of $j, k, m, n, p,$ q, r and t is $\dfrac{1+2+3+4+5+6+7+28}{8} = \dfrac{56}{8} = 7$. Put a nice big, dark circle around this number. Now plug $t = 28$ into each answer choice.

    A.   14
    B. ≈ 9.33
    C.   7
    D.   5.6
    E. ≈ 4.67

Since A, B, D and E are incorrect we can eliminate them. Therefore, the answer is choice **C**.

Important note: C is **not** the correct answer simply because it is equal to 7. It is correct because all four of the other choices are **not** 7.

156. A group of students take a test and the average score is 65. One more student takes the test and receives a score of 92 increasing the average score of the group to 68. How many students were in the initial group?

- **F.** 5
- **G.** 6
- **H.** 7
- **J.** 8
- **K.** 9

* **Solution by changing averages to sums:** Let n be the number of students in the initial group. We change the average to a sum using the formula

$$\text{Sum} = \text{Average} \cdot \text{Number}$$

So the initial **Sum** is $65n$.

When we take into account the new student, we can find the new sum in two different ways.

(1) We can add the new score to the old sum to get $65n + 92$.

(2) We can compute the new sum directly using the simple formula above to get $68(n + 1) = 68n + 68$.

We now set these equal to each other and solve for n:

$$65n + 92 = 68n + 68$$
$$24 = 3n$$
$$n = 8.$$

This is choice **J.**

LEVEL 5: TRIGONOMETRY

157. In the right triangle below, $b > a > 0$. One of the angle measures in the triangle is $\sin^{-1}\frac{b}{\sqrt{a^2+b^2}}$. What is $\tan[\sin^{(-1)}\left(\frac{b}{\sqrt{a^2+b^2}}\right)]$?

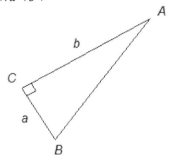

A. $\dfrac{b}{a}$

B. $\dfrac{a}{b}$

C. $\dfrac{b}{\sqrt{a^2+b^2}}$

D. $\dfrac{a}{\sqrt{a^2+b^2}}$

E. $\dfrac{\sqrt{a^2+b^2}}{a}$

* First note that by the Pythagorean Theorem, $AB = \sqrt{a^2 + b^2}$. Since $\sin B = \frac{\text{OPP}}{\text{HYP}} = \frac{b}{\sqrt{a^2+b^2}}$, it follows that $\sin^{-1}\frac{b}{\sqrt{a^2+b^2}} = B$. So we have $\tan[\sin^{(-1)}\left(\frac{b}{\sqrt{a^2+b^2}}\right)] = \tan B = \frac{\text{OPP}}{\text{ADJ}} = \frac{b}{a}$, choice **A**.

Notes: (1) $\sin^{-1} x = y$ is essentially the same as $\sin y = x$.

To compute $\sin y$ we input an angle and get out a number. To compute $\sin^{-1} x$ we input a number and get out an angle.

(2) We can get the answer to this problem in just a few seconds without any writing. Here is how to think about it. For the expression $\tan[\sin^{(-1)}\left(\frac{b}{\sqrt{a^2+b^2}}\right)]$, we are simply computing the tangent of an angle. We just need to figure out if we want angle A or angle B.

We choose the correct angle by looking at the sine of each angle. We want the sine of the angle to be $\frac{b}{\sqrt{a^2+b^2}}$. In other words, we just want the opposite side to the angle to be b. Well this is angle B. So we just need to compute $\tan B$.

To summarize, since the side opposite B has length b, we simply compute $\tan B = \frac{b}{a}$.

158. For $0 < x < \frac{\pi}{2}$, the expression $\frac{\sin x}{\sqrt{1-\sin^2 x}} - \frac{\sqrt{1-\sin^2 x}}{\cos x}$ is equivalent to:

    F. $\tan x$
    G. $\cot x$
    H. $\tan x - 1$
    J. $1 - \tan x$
    K. $1 - \cot x$

* Since $1 - \sin^2 x = \cos^2 x$, we have

$$\frac{\sin x}{\sqrt{1-\sin^2 x}} - \frac{\sqrt{1-\sin^2 x}}{\cos x} = \frac{\sin x}{\sqrt{\cos^2 x}} - \frac{\sqrt{\cos^2 x}}{\cos x} = \frac{\sin x}{\cos x} - \frac{\cos x}{\cos x} = \tan x - 1$$

This is choice **H.**

Note: (1) One of the most important trigonometric identities is the Pythagorean Identity which says

$$\cos^2 x + \sin^2 x = 1.$$

From this we also get the equations

$$\sin^2 x = 1 - \cos^2 x \quad \text{and} \quad \cos^2 x = 1 - \sin^2 x$$

by performing a simple subtraction.

(2) Since $0 < x < \frac{\pi}{2}$, $\cos x$ is positive. So $\sqrt{\cos^2 x} = \cos x$. If we didn't have a restriction on x, then we could only say $\sqrt{\cos^2 x} = |\cos x|$.

(3) We also needed the simple Quotient Identity $\frac{\sin x}{\cos x} = \tan x$.

(4) This problem could also be solved by picking numbers. Such a solution is done in detail for problem 125.

159. A ladder rests against the side of a wall and reaches a point that is 25 meters above the ground. The angle formed by the ladder and the ground is 58°. A point on the ladder is 3 meters from the wall. What is the vertical distance, in meters, from this point on the ladder to the ground?

   A. $22 \tan 58°$
   B. $22 \cos 58°$
   C. $25 - 3 \sin 58°$
   D. $25 - 3 \cos 58°$
   E. $25 - 3 \tan 58°$

* Let's draw a picture.

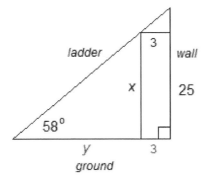

Note that there are two triangles in this picture. We will need to use both of them.

Also recall that for any angle A, $\tan A = \frac{\text{OPP}}{\text{ADJ}}$ (see the end of the solution to problem 61). Using the smaller triangle we have $\tan 58° = \frac{x}{y}$ and using the larger triangle we have $\tan 58° = \frac{25}{y+3}$. The first equation gives $y \tan 58° = x$. and the second equation gives $(y + 3) \tan 58° = 25$. Distributing this last equation on the left gives $y \tan 58° + 3 \tan 58° = 25$. Substituting from the first equation yields $x + 3 \tan 58° = 25$. We subtract $3 \tan 58°$ from each side of this last equation to get $x = 25 - 3 \tan 58°$, choice **E**.

160. Triangle *PQR* is shown in the figure below. The measure of ∠*P* is 32°, *PQ* = 9 in, and *PR* = 15 in. Which of the following is the length, in inches, of $\overline{QR}$?

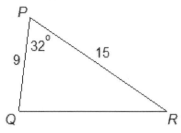

F. $9\sin 32°$

G. $15\sin 32°$

H. $\sqrt{15^2 - 9^2}$

J. $\sqrt{15^2 + 9^2}$

K. $\sqrt{15^2 + 9^2 - 2(15)(9)\cos 32°}$

* We use the law of cosines to get

$$QR^2 = 15^2 + 9^2 - 2(15)(9)\cos 32°$$

So $QR = \sqrt{15^2 + 9^2 - 2(15)(9)\cos 32°}$, choice **K.**

Notes: (1) The **law of cosines** says $c^2 = a^2 + b^2 - 2ab\cos C$ where a, b, and c are the lengths of the sides of the triangle, and the side of length c is opposite angle C.

(2) This problem gives the most direct use of the law of cosines. In this example two side lengths of a triangle and the angle between these two sides are known. We are trying to find the length of the side opposite the known angle.

(3) Observe that QR is by itself on one side of the equation because it is opposite angle P.

(4) It is not actually necessary to know the formula for the Law of Cosines in detail to get the answer to this problem. A vague notion of what it looks like should get you the correct answer.

Solution by process of elimination: The first four answer choices all involve computations that generally require a right triangle. Since the given triangle is <u>not</u> a right triangle, it is unlikely that any of them are the answer. So a reasonable guess is choice **K.**

Note: Choices **F** and **G** look like they are computations from basic right triangle trigonometry, while choices **H** and **J** look like straightforward applications of the Pythagorean Theorem.

SUPPLEMENTAL PROBLEMS

Full solutions to these problems are available for free download here:

www.thesatmathprep.com/ACTprmX.html

LEVEL 1: NUMBER THEORY

1. $|8 - 2| - |2 - 8| = ?$
    A. -12
    B. -4
    C. 0
    D. 4
    E. 12

2. The fourth term of an arithmetic sequence is 19 and the fifth term is 13. What is the third term?
    F. -21
    G. -6
    H. 6
    J. 21
    K. 25

3. To keep up with inflation, a store owner raises the price of a $30 item by 26%. What is the new price of the item?
    A. $30.22
    B. $32.20
    C. $37.00
    D. $37.80
    E. $52.00

4. On Monday Jason spent one third of his allowance. On Tuesday he spent one third of the remaining money, and on Wednesday he spent one third of what remained from Tuesday. If $8 then remained, how much did he originally receive for his allowance?
    F. $12
    G. $18
    H. $27
    J. $54
    K. $96

147

5. You are about to pay for a hat priced at $9.99. A sales tax of 9% of $9.99 will be added (rounded to the nearest cent). You have 15 one-dollar bills, but how much will you need, in cents, if you want to be ready with exact change?

   A. 89
   B. 75
   C. 53
   D. 41
   E. 33

6. Which of the following is <u>not</u> a factor of 2431 ?

   F. 1
   G. 7
   H. 11
   J. 13
   K. 17

7. What is the smallest integer greater than $\sqrt{50}$?

   A. 6
   B. 7
   C. 8
   D. 9
   E. 10

8. On the first Monday in February, Mrs. Green gave her students 7 math problems to solve. On each school day after that, she gave the students 8 math problems to solve. During the first 15 school days, how many math problems had she given the students to solve?

   F. 15
   G. 100
   H. 105
   J. 119
   K. 120

148

LEVEL 1: ALGEBRA AND FUNCTIONS

9.   Which of the following mathematical expressions is equivalent to the verbal expression "A number, c, squared is 52 more than the product of c and 11"?

   A. $2c = 52 + 11c$
   B. $2c = 52c + 11c$
   C. $c^2 = 52 - 11c$
   D. $c^2 = 52 + c^{11}$
   E. $c^2 = 52 + 11c$

10.   If $9 + 5x = 39$, then $4x = ?$

   F.   6
   G.   12
   H.   18
   J.   24
   K.   30

11.   If $a = 4$, $b = 3$, and $c = -7$, then what is the value of $(b + c)(a + b - c)$?

   A.   −56
   B.   − 4
   C.   0
   D.   4
   E.   56

12.   For what value of x is the equation $x + 5(x - 4) = 4$ true?

   F.   24
   G.   20
   H.   8
   J.   4
   K.   3

13.   The expression $9ab - 2a(4a + 5b)$ is equivalent to:

   A. $-ab - 8a^2$
   B. $2ab - 6a$
   C. $10ab - 8a^2$
   D. $-9ab$
   E. $-8a^2$

149

14. What is the value of the expression $(b - a)^2$ when $a = -1$ and $b = 6$?

    F.   4
    G.   9
    H. 25
    J.   36
    K. 49

15. If $A = -5x$ and $B = 3y - x$, then what is the value of $A - B$?

    A. $-6x - 3y$
    B. $-6x + 3y$
    C. $-4x - 3y$
    D. $-4x + 3y$
    E. $4x - 3y$

16. When written in symbols, "The square of the sum of a and b" is represented as:

    F. $(a + b)^2$
    G. $a^2 + b$
    H. $a + b^2$
    J. $a^2 + b^2$
    K. $(a^2 + b^2)^2$

LEVEL 1: GEOMETRY

17. In isosceles triangle $\triangle CAT$, $\angle C$ and $\angle A$ are congruent and the measure of $\angle T$ is 52°. What is the measure of $\angle C$?

    A. 148°
    B. 128°
    C. 102°
    D.   86°
    E.   64°

18. What is the area, in feet, of a rectangle with length 3 ft and width 15 ft?

    F. 18
    G. 21
    H. 36
    J. 45
    K. 90

19. In rectangle *PQRS*, which of the following must be true about the measures of ∠*PQR* and ∠*QRS* ?

   A. each are 90°
   B. each are less than 90°
   C. each are greater than 90°
   D. they add up to 90°
   E. they add up to 360°

20. A point at (6,–5) in the standard (*x*, *y*) coordinate plane is reflected in the *x*-axis. What are the coordinates of the new point?

   F. (–6, –5)
   G. (–6, 5)
   H. (6, 5)
   J. (–5, 6)
   K. (5, –6)

21. The interior dimensions of a cube are 5 inches by 5 inches by 5 inches. What is the volume, in cubic inches, of the interior of the cube?

   A. 15
   B. 30
   C. 120
   D. 125
   E. 150

22. In the standard (*x*, *y*) coordinate plane, point *M* with coordinates (3,7) is the midpoint of $\overline{PQ}$, and *P* has coordinates (1,9). What are the coordinates of *Q* ?

   F. (5, 5)
   G. (–5, –5)
   H. (–1, 11)
   J. (2, 8)
   K. (4, 16)

151

23. In the isosceles right triangle below, $PQ = 7$ inches. What is the length, in inches, of $\overline{PR}$?

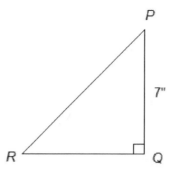

   A. $7\sqrt{2}$
   B. $\sqrt{14}$
   C. 14
   D. 7
   E. 3.5

24. In the figure below, adjacent sides meet at right angles and the lengths given are in inches. What is the perimeter of the figure, in inches?

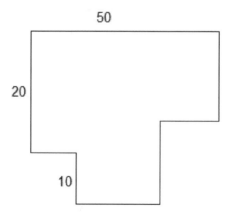

   F.   80
   G. 100
   H. 110
   J.   160
   K. 1500

LEVEL 1: PROBABILITY AND STATISTICS

25. The daily totals of shoppers in the Moonlight Convenience Store last week were 97, 123, 216, 26, 41, 187, and 87. What was the average number of shoppers each day?

   A.   60
   B.   111
   C.   155
   D.   259
   E.   775

26. A menu lists 3 appetizers, 4 meals, 2 drinks, and 2 desserts. A dinner consists of 1 of each of these 4 items. How many different dinners are possible from this menu?

   F.   2
   G.   12
   H.   48
   J.   72
   K.   120

27. A 20-member committee needs to choose a treasurer. They decide that the treasurer, who will be chosen at random, CANNOT be the president or vice president of the committee. What is the probability that Jessie, who is a member of the committee, will be chosen? (Assume that Jessie is not the president or vice president of the committee.)

   A.   0
   B. $\frac{1}{3}$
   C. $\frac{1}{9}$
   D. $\frac{1}{18}$
   E. $\frac{1}{20}$

153

28. A data set contains 7 elements and has a mean of 8. Six of the elements are 1, 2, 3, 5, 8 and 20. Which of the following is the seventh element?

   F.   6
   G.   9
   H.   10
   J.   15
   K.   17

29. If the probability that it will be cloudy tomorrow is 0.9, what is the probability that it will <u>not</u> be cloudy tomorrow?

   A.   1.4
   B.   1.0
   C.   0.6
   D.   0.1
   E.   0.0

30. The average of 10 numbers is 1.3. If each of the numbers is increased by 6, what is the average of the 10 new numbers?

   F.   0.0
   G.   0.3
   H.   1.3
   J.   2.3
   K.   7.3

31. Of the 30 students in a math class, 7 earned A's, 12 earned B's, 5 earned C's, and 6 earned D's. If a student from the class is chosen at random, what is the probability that the student chosen had earned a B in the class?

   A. $\dfrac{1}{5}$

   B. $\dfrac{2}{5}$

   C. $\dfrac{3}{5}$

   D. $\dfrac{2}{3}$

   E. $\dfrac{4}{5}$

154

32. In a math class, a student's overall grade for the semester is determined by throwing out the lowest test grade and taking the average of the remaining test grades. Alena received test grades of 67, 71, 83, 91, 95, and 98 this semester. What overall grade did Alena receive in the math class this semester?

    F. 82
    G. 84.2
    H. 87
    J. 87.6
    K. 92

LEVEL 2: NUMBER THEORY

33. The expression $\dfrac{\frac{1}{3}+3}{1+\frac{1}{5}}$ is equal to:

    A. $\dfrac{23}{4}$
    B. $\dfrac{25}{9}$
    C. 3
    D. 2
    E. 1

34. What is the least common denominator for adding the fractions $\dfrac{3}{14}, \dfrac{1}{18}$, and $\dfrac{5}{27}$?

    F. 1
    G. 42
    H. 54
    J. 378
    K. 756

35. Which of the following is equivalent to $8^{1/4}$?

    A. -1×8^4
    B. $\sqrt[4]{8}$
    C. $\sqrt{4}$
    D. $\dfrac{1}{8^4}$
    E. 4

155

36. The odometer in Don's car read 42,926 miles when he left on vacation and 43,406 miles when he returned. Don drove his car for a total of 10 hours during the trip. Based on these odometer readings, what was Don's average driving speed for the duration of the trip?

    F. 64
    G. 58
    H. 52
    J. 48
    K. 40

37. John purchased a house for $340,000. He financed all of the $340,000 and started loan payments of $2100 a month for 30 years. At the end of the 30-year period, how much more than the purchase price will John have paid for his house?

    A. $525,000
    B. $416,000
    C. $340,000
    D. $160,000
    E. $102,250

38. For two consecutive integers, the result of adding the larger integer and twice the smaller integer is 106. What is the larger integer?

    F. 12
    G. 25
    H. 35
    J. 36
    K. 72

39. For all positive integers m, what is the greatest common factor of the 3 numbers $40m$, $60m$, and $100m$?

    A. 20
    B. 40
    C. m
    D. $20m$
    E. $40m$

156

40. What is the correct ordering of $\sqrt{2}$, 1.4, and $\frac{\sqrt{5}}{2}$ from least to greatest?

F. $\sqrt{2} < 1.4 < \frac{\sqrt{5}}{2}$

G. $1.4 < \sqrt{2} < \frac{\sqrt{5}}{2}$

H. $\sqrt{2} < \frac{\sqrt{5}}{2} < 1.4$

J. $1.4 < \frac{\sqrt{5}}{2} < \sqrt{2}$

K. $\frac{\sqrt{5}}{2} < 1.4 < \sqrt{2}$

LEVEL 2: ALGEBRA AND FUNCTIONS

41. For the function $f(x) = 5x^2 - 7x$, what is the value of $f(-3)$?

A. −66
B. 24
C. 24
D. 66
E. 246

42. $(x - 2y + 3z) - (5x - 4y + 6z)$ is equivalent to:

F. $-6x - 6y + 9z$
G. $-6x + 2y - 3z$
H. $-4x - 6y - 3z$
J. $-4x + 2y + 9z$
K. $-4x + 2y - 3z$

43. $(y^7)^4$ is equivalent to:

A. y^{28}
B. y^{11}
C. $4y^8$
D. $4y^7$
E. $28y$

157

44. Let a function of 2 variables be defined by $g(x, y) = xy + 3xy^2 - (x - y^2)$, what is the value of $g(2, -1)$?

    F. 1
    G. 2
    H. 3
    J. 4
    K. 5

45. For which nonnegative value of b is the expression $\frac{1}{2-b^2}$ undefined?

    A. 0
    B. 2
    C. 8
    D. $\sqrt{2}$
    E. $\sqrt{8}$

46. If $\frac{1}{5} \geq \frac{7}{x}$, what is the smallest possible positive value for x ?

    F. $\frac{1}{5}$
    G. 3
    H. 17
    J. 35
    K. 70

47. For all x, $7 - 5(2 - x) = ?$

    A. $-5x + 17$
    B. $-5x - 1$
    C. $-5x - 3$
    D. $5x - 1$
    E. $5x - 3$

48. If $3x^2 + 9x = 84$, what are the possible values for x ?

    F. -4 and 7
    G. -7 and 4
    H. -7 and -4
    J. -7 and -12
    K. 12 and 14

LEVEL 2: GEOMETRY

49. One side of square *PQRS* is 24 feet long. A rectangle with the same area as square *PQRS* has a width of 48 feet. What is the rectangle's length, in feet?

    A. 12
    B. 16
    C. 20
    D. 48
    E. 108

50. What is the slope of any line parallel to the line $3x - 2y = 5$?

    F. -2
    G. $-\dfrac{3}{5}$
    H. $\dfrac{3}{2}$
    J. 3
    K. 5

51. For all triangles $\triangle ABC$ where the measure of angle B is greater than the measure of angle A, such as the triangle shown below, which of the following statements is true?

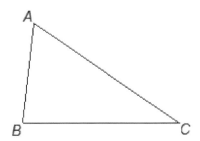

    A. Sometimes $AC < BC$ and sometimes $AC = BC$
    B. Sometimes $AC > BC$ and sometimes $AC = BC$
    C. $AC < BC$
    D. $AC = BC$
    E. $AC > BC$

52. In the standard (x, y) coordinate plane, what is the slope of the line joining the points $(-5, 4)$ and $(3, -2)$?

F. $-\frac{4}{3}$

G. $-\frac{3}{4}$

H. $\frac{1}{3}$

J. $\frac{3}{4}$

K. $\frac{4}{3}$

53. What is the perimeter of polygon *PQRST* shown below, in inches?

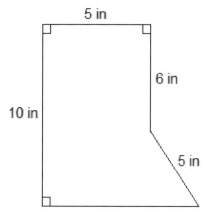

A. 34
B. 36
C. 38
D. 40
E. 42

54. An isosceles right triangle has a hypotenuse with length $5\sqrt{2}$ meters. What is the length of one of the legs, in meters?

F. $\sqrt{10}$
G. $\sqrt{20}$
H. 5
J. 10
K. 20

160

55. A rectangular room that is 3 meters longer than it is wide has an area of 154 square meters. What is the perimeter of the room?

   A. 11
   B. 14
   C. 22
   D. 28
   E. 50

56. If point W has a nonzero x-coordinate and a nonzero y-coordinate and the coordinates have the same sign, then point W <u>must</u> be located in which of the 4 quadrants?

   F. I only
   G. II only
   H. III only
   J. I or III only
   K. II or IV only

LEVEL 2: PROBABILITY AND STATISTICS

57. What is the median of the following 9 test grades?

   95, 72, 81, 96, 62, 98, 82, 76, 82

   A. 81
   B. 82
   C. 82.6
   D. 91
   E. 95

58. Three light bulbs are placed into three different lamps. How many different arrangements are possible for three light bulbs of different colors – one white, one black, and one yellow?

   F. 1
   G. 3
   H. 4
   J. 6
   K. 27

59. The average (arithmetic mean) of 22, 50, and y is 50. What is the value of y ?

    A. 134
    B.   78
    C.   76
    D.   72
    E.   50

60. Of the marbles in a jar, 52 are white. William randomly takes one marble out of the jar. If the probability is $\frac{13}{15}$ that the marble he chooses is white, how many marbles are in the jar?

    F. 45
    G. 52
    H. 56
    J. 58
    K. 60

LEVEL 2: TRIGONOMETRY

61. For $\angle P$ in $\triangle PQR$ below, which of the following trigonometric expressions has value $\frac{13}{5}$?

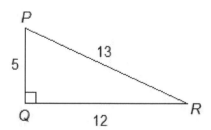

    A. tan P
    B. cot P
    C. sin P
    D. cos P
    E. sec P

62. The figure below shows a right triangle whose hypotenuse is 7 feet long. How many feet long is the longer leg of this triangle?

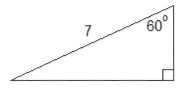

   F.   3.5

   G. 14

   H. $\dfrac{7\sqrt{3}}{2}$

   J. $\dfrac{7\sqrt{3}}{6}$

   K. $\dfrac{14\sqrt{3}}{3}$

63. The dimensions of the right triangle below are given in meters. What is tan B ?

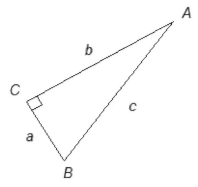

   A. $\dfrac{c}{b}$

   B. $\dfrac{a}{b}$

   C. $\dfrac{a}{c}$

   D. $\dfrac{b}{a}$

   E. $\dfrac{b}{c}$

64. As shown below, a 14-foot ramp forms an angle of 63° with the vertical wall it is leaning against. Which of the following is an expression for the horizontal length, in feet, of the ramp?

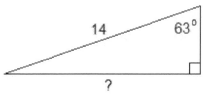

F. 14 cos 63°

G. 14 sin 63°

H. 14 tan 63°

J. 14 cot 63°

K. 14 sec 63°

LEVEL 3: NUMBER THEORY

65. The number line graph below is the graph of which of the following inequalities?

A. −3 ≤ *x* and 5 ≤ *x*

B. −3 ≤ *x* and 5 ≥ *x*

C. −3 ≤ *x* or 5 ≤ *x*

D. −3 ≥ *x* or 5 ≤ *x*

E. −3 ≥ *x* or 5 ≥ *x*

66. The ratio of the number of boys to the number of girls in a park is 3 to 5. What percent of the children in the park are girls?

F. 12.5%

G. 37.5%

H. 60%

J. 62.5%

K. 70%

67. Which of the following is not a rational number?

    A. $\sqrt{3^2}$

    B. $\dfrac{\sqrt{\pi^2}}{\pi}$

    C. $(\sqrt{11})^2$

    D. $\sqrt{\dfrac{16}{49}}$

    E. $\sqrt{\dfrac{\pi^2}{9}}$

68. A recipe for a sports drink calls for 11 parts fruit juice to 5 parts water. To make 64 quarts of this drink, how many quarts of water should be used?

    F. 4
    G. 16
    H. 20
    J. 22
    K. 44

69. If 20% of a given number is 16, then what is 30% of the given number?

    A. 8
    B. 12
    C. 24
    D. 48
    E. 80

70. Aaron, Bert, and Carlos shared a pumpkin pie. Aaron ate $\dfrac{1}{8}$ of the pie, Bert ate $\dfrac{2}{3}$ of the pie, and Carlos ate the rest. What is the ratio of Carlos's share to Bert's share to Aaron's share?

    F. 24:6:8
    G. 3:16:5
    H. 3:24:16
    J. 5:16:3
    K. 5:16:24

165

71. Alex plans to drive 550 miles from New York to Virginia, driving an average of 50 miles per hour. How many miles per hour faster must he average, while driving, to reduce his total driving time by 1 hour?

   A. 20
   B. 18
   C. 15
   D. 8
   E. 5

72. What is the least common multiple of the numbers 2, 3, 4, 5, 6, and 7?

   F. 1
   G. 105
   H. 210
   J. 420
   K. 5040

LEVEL 3: ALGEBRA AND FUNCTIONS

73. The expression $(3b - 2)(b + 5)$ is equivalent to:

   A. $3b^2 - 7$
   B. $3b^2 - 10$
   C. $3b^2 - 2b - 7$
   D. $3b^2 + 13b - 10$
   E. $3b^2 - 13b - 10$

74. For all x, $(x^2 - 3x + 1)(x + 2) = ?$

   F. $x^3 - x^2 - 5x + 2$
   G. $x^3 - x^2 - 5x - 2$
   H. $x^3 - x^2 + 5x + 2$
   J. $x^3 + x^2 - 5x + 2$
   K. $x^3 + x^2 - 5x - 2$

75. Which of the following values of x is in the solution set of $x^2 + 5x - 6 = 8$?

   A. -14
   B. -7
   C. -6
   D. -2
   E. 1

76. The value of x that will make $\frac{x}{3} - 2 = -\frac{11}{4}$ a true statement lies between which of the following numbers?

   F. -3 and -2
   G. -2 and -1
   H. -1 and 0
   J. 0 and 1
   K. 1 and 2

77. For all x and y, $(3y - x)(y^3 + x^2) = ?$

   A. $3y^3 - x^3$
   B. $3y^4 - x^3$
   C. $3y^4 + 2x^2y - x^3$
   D. $3y^4 + 3x^2y - x^3$
   E. $3y^4 + 3x^2y - xy^3 - x^3$

78. The operation ■ is defined as $a\ ■\ b = \frac{2b^2 - 8a^2}{b + 2a}$ where a and b are real numbers and $b \neq -2a$. What is the value of $(-2)\ ■\ (-1)$?

   F. 14
   G. 12
   H. 6
   J. -2
   K. -6

79. The inequality $5(x - 3) > 6(x - 2)$ is equivalent to which of the following inequalities?

   A. $x < -3$
   B. $x > -3$
   C. $x < 6$
   D. $x > 6$
   E. $x < 12$

167

80. It costs $(s + t)$ dollars for a box of brand A cat food, and $(q - r)$ dollars for a box of brand B cat food. The difference between the cost of 15 boxes of brand A cat food and 7 boxes of brand B cat food is k dollars. Which of the following equations represents a relationship between s, t, q, r, and k?

   F. $105(s + t)(q - r) = k$
   G. $|7(q - r) + 15(s - t)| = k$
   H. $|15(s + t) + 7(q - r)| = k$
   J. $|15(s + t) - 7(q - r)| = k$
   K. $\frac{15(s+t)}{7(q-r)} = k$

LEVEL 3: GEOMETRY

81. What is the slope of any line parallel to the y-axis in the (x, y) coordinate plane?

   A. -1
   B. 0
   C. 1
   D. Undefined
   E. Cannot be determined from the given information

82. A circle in the standard (x, y) coordinate plane has equation $(x - 5)^2 + (y + 2)^2 = 73$. What are the radius of the circle, in coordinate units, and the coordinates of the center of the circle?

   F. $\sqrt{73}$, $(-5, \ 2)$
   G. $\sqrt{73}$, $(\ 5, -2)$
   H. $\sqrt{73}$, $(-2, \ 5)$
   J. 73, $(\ 2, -5)$
   K. 73, $(-5, \ 2)$

83. For what value of k would the following system of equations have no solutions?
$$5kx - 4y = 4$$
$$10x - 2y = 3$$

   A. -10
   B. -4
   C. 2
   D. 4
   E. 10

168

84. Which of the following figures in a plane separates it into half-planes?

   F. A point
   G. An angle
   H. A line segment
   J. A ray
   K. A line

85. In the standard (x, y) coordinate plane, which of the following lines passes through the point $(0, -2)$ and is perpendicular to the line $y = -3x + 5$?

   A. $y = -3x - 2$

   B. $y = -\frac{1}{3}x - 2$

   C. $y = \frac{1}{3}x - 2$

   D. $y = -\frac{1}{3}x + 2$

   E. $y = \frac{1}{3}x + 2$

86. In the figure below, where $\triangle CAT \sim \triangle DOG$, lengths given are in inches. What is the perimeter, in inches, of $\triangle DOG$?

   (Note: The symbol $\sim$ means "is similar to.")

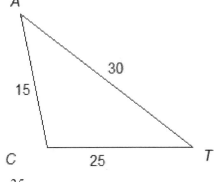

   F. 35
   G. 28
   H. 21
   J. 14
   K. 7

87. The radius of a circle is $\frac{7}{\sqrt{\pi}}$ centimeters. How many centimeters long is its circumference?

    A. $\frac{14}{\sqrt{\pi}}$

    B. $\frac{14}{\pi}$

    C. $\frac{14}{\pi^2}$

    D. 14

    E. $14\sqrt{\pi}$

88. Point A is a vertex of a 6-sided polygon. The polygon has 6 sides of equal length and 6 angles of equal measure. When all possible diagonals are drawn from point A in the polygon, how many triangles are formed?

    F. One
    G. Two
    H. Three
    J. Four
    K. Five

LEVEL 3: PROBABILITY AND STATISTICS

89. Jeff has taken 6 of 10 equally weighted math tests this semester, and he has an average score of exactly 82.0 points. How many points does he need to earn on the 7th test to bring his average score up to exactly 83 points?

    A. 86
    B. 87
    C. 88
    D. 89
    E. 90

170

90. Jenny has red, black and white marbles in a bag. There are 4 more red marbles than black, 8 more black marbles than white, and 80 marbles altogether. If Jenny chooses one marble from the bag at random, what is the probability that she will choose a white marble?

F. $\frac{1}{6}$

G. $\frac{1}{5}$

H. $\frac{1}{4}$

J. $\frac{1}{3}$

K. $\frac{1}{2}$

91. The test grades of the 26 students in a math class are shown in the chart below. What is the median test grade for the class?

TEST GRADES OF STUDENTS IN MATH CLASS

Test Grade	75	82	87	93	100
Number of students with that grade	5	7	10	3	1

A. 75
B. 82
C. 87
D. 90
E. 93

92. A wall is to be painted one color with a stripe of a different color running through the middle. If 7 different colors are available, how many color combinations are possible?

F. 42
G. 21
H. 14
J. 13
K. 7

171

LEVEL 3: TRIGONOMETRY

93. In right triangle PQR below, the measure of $\angle R$ is 90°, $PR = 2$ units, and $RQ = 3$ units. What is sin P ?

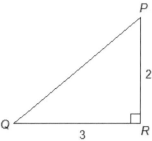

A. $\dfrac{2}{3}$

B. $\dfrac{3}{2}$

C. $\dfrac{\sqrt{13}}{2}$

D. $\dfrac{2}{\sqrt{13}}$

E. $\dfrac{3}{\sqrt{13}}$

94. The 2 triangles in the figure below share a common side. What is cos $(x + y)$?

(Note: cos $(x + y) = \cos x \cos y - \sin x \sin y$ for all x and y.)

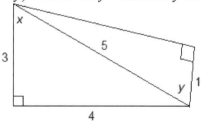

F. $\dfrac{4}{5}$

G. $\dfrac{3+4\sqrt{24}}{5}$

H. $\dfrac{4\sqrt{24} - 3}{25}$

J. $\dfrac{3-4\sqrt{24}}{25}$

K. $\dfrac{27}{25}$

95. A dog, a cat, and a mouse are all sitting in a room. Their relative positions to each other are described in the figure below. Which of the following expressions gives the distance, in feet, from the cat to the dog?

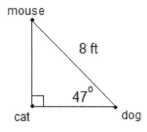

A. $8 \sin 47°$

B. $8 \cos 47°$

C. $8 \tan 47°$

D. $\dfrac{8}{\sin 47°}$

E. $\dfrac{8}{\cos 47°}$

96. In the figure below, $\tan k = ?$

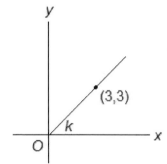

F. $3\sqrt{2}$

G. $\dfrac{3\sqrt{2}}{2}$

H. $\dfrac{\sqrt{3}}{2}$

J. 1

K. $\dfrac{1}{3}$

LEVEL 4: NUMBER THEORY

97. What is the value of $\log_3 81$?

   A. 3
   B. 4
   C. 8
   D. 9
   E. 27

98. Which of the following calculations will yield twice an odd integer for any integer?

   F. $2k$
   G. $2k + 1$
   H. $4k$
   J. $4k + 1$
   K. $4k + 2$

99. What is the least positive integer that gives a remainder of 3 when divided by 5 and 7?

   A. 8
   B. 10
   C. 35
   D. 38
   E. 73

100. Two numbers are reciprocals if their product is equal to 1. If a and b are <u>negative</u> reciprocals and $a < -1$, then b must be:

   F. greater than 1
   G. between 0 and 1
   H. equal to 0
   J. between -1 and 0
   K. less than -1

101. The ratio of j to k is 7 to 1, and the ratio of k to n is 1 to 3. What is the value of $\frac{2j+5k}{4k+3n}$?

   A. $\frac{1}{3}$
   B. $\frac{13}{19}$
   C. 1
   D. $\frac{19}{13}$
   E. 3

102. The 517th digit after the decimal point in the repeating decimal $0.\overline{784398}$ is

   F. 3
   G. 4
   H. 7
   J. 8
   K. 9

103. $\dfrac{1}{5} \cdot \dfrac{3}{6} \cdot \dfrac{4}{7} \cdot \dfrac{5}{8} \cdot \dfrac{6}{9} \cdot \dfrac{7}{10} \cdot \dfrac{8}{11} \cdot \dfrac{9}{12} = ?$

   A. $\dfrac{1}{110}$

   B. $\dfrac{1}{55}$

   C. $\dfrac{1}{12}$

   D. 1

   E. 3

104. Jason cut a piece of paper into 5 equal pieces. He threw one piece away, and cut each of the remaining pieces into 4 equal pieces. He again threw one of these pieces away, and cut each of the remaining pieces into 3 equal pieces. He again threw one piece away and cut each of the remaining pieces into 2 equal pieces. After throwing 1 more piece away, how many pieces of paper does Jason have left in total?

   F. 43
   G. 44
   H. 86
   J. 87
   K. 88

LEVEL 4: ALGEBRA AND FUNCTIONS

105. The *determinant* of a matrix $\begin{bmatrix} a & b \\ c & d \end{bmatrix}$ is equal to $ad - bc$. If
$\begin{bmatrix} x - y & 3 \\ 3 & x - y \end{bmatrix} = k$, then what must k be equal to in order for
$x = y$ to be true?

   A. -12
   B. -9
   C. 0
   D. 9
   E. 12

106. Which of the following has the least value for all values of x in the interval $-1 < x < 0$?

   F. x^3
   G. x^2
   H. $-|x|$
   J. $-\sqrt{-x}$
   K. $-x$

107. If $3a^5 b^4 < 0$, then which of the following CANNOT be true?

   A. $a < 0$

   B. $a > 0$

   C. $b < 0$

   D. $b > 0$

   E. $b = a$

108. If $\dfrac{x^a x^b}{(x^c)^d} = x^2$ for all $x \neq 0$, which of the following must be true?

   F. $a + b - cd = 2$

   G. $\dfrac{a+b}{cd} = 2$

   H. $ab - cd = 2$

   J. $ab - c^d = 2$

   K. $\dfrac{ab}{c^d} = 2$

109. After solving a quadratic equation by completing the square, it was found that the equation had solutions, $x = -3 \pm \sqrt{-15b^2}$ where b is a positive real number. Which of the following gives the solutions as complex numbers?

   A. $-3 \pm \quad bi$
   B. $-3 \pm \sqrt{15}bi$
   C. $-3 \pm \quad 5bi$
   D. $-3 \pm 15bi$
   E. $-3 \pm 225bi$

110. Let $a \odot b = (-a^2 + b)^3$ for all integers a and b. Which of the following is the value of $-1 \odot 2$?

   F. -27
   G. -1
   H. 0
   J. 1
   K. 27

111. If x and y are real numbers such that $x > 2$ and $y < 0$, then which of the following inequalities <u>must</u> be true?

   A. $\frac{x}{y} > 2$

   B. $x < y^2$

   C. $\frac{x}{7} - 8 > \frac{y}{7} - 8$

   D. $x^2 > y^2$

   E. $\frac{1}{x^2} > \frac{1}{y^2}$

112. If $c > 0$, $s^2 + t^2 = c$, and $st = c + 5$, what is $(s + t)^2$ in terms of c ?

   F. $c + 5$
   G. $c + 10$
   H. $2c + 5$
   J. $2c + 10$
   K. $3c + 10$

LEVEL 4: GEOMETRY

113. The equations below are linear equations of a system where m, n, and k are positive integers.

$$mx - ny = k$$
$$mx + ny = k$$

Which of the following describes the graph of at least 1 such system of equations in the standard (x, y) coordinate plane?

      I. A single line
      II. 2 intersecting lines
      III. 2 parallel lines

  A. I only
  B. II only
  C. III only
  D. I or II only
  E. I, II, or III

114. The edges of a cube are each 4 cm long. What is the surface area, in cm^2, of this cube?

  F. 4
  G. 16
  H. 64
  J. 96
  K. 128

115. A circle with a radius of 5 centimeters is inscribed in square $PQRS$, as shown below. What is the area, in square centimeters, of the shaded region?

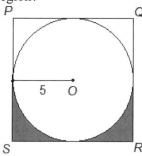

  A. $100 - 25\pi$
  B. $100 - \dfrac{25\pi}{2}$
  C. $50 - 25\pi$
  D. $50 - \dfrac{25\pi}{2}$
  E. $5\pi - 10$

178

116. In the (x, y) coordinate plane, the vertices of ΔPQR are $(8,10)$, $(12,6)$, and $(4,6)$, respectively. What is the perimeter of ΔPQR ?

   F. 14
   G. 20
   H. $8 + 4\sqrt{2}$
   J. $8 + 8\sqrt{2}$
   K. 28

117. In pentagon *HOUSE* below, $\angle H$ measures 50°. The measures of the other 4 angles are equal. What is the measure of $\angle U$?

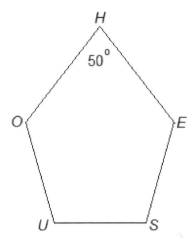

   A. 47°
   B. 47.5°
   C. 122.5°
   D. 245°
   E. 490°

118. In the (x, y) coordinate plane, which of the following is an equation of the circle having the points $(3,-3)$ and $(-5,5)$ as endpoints of a diameter?

   F. $(x + 1)^2 + (y - 1)^2 = \sqrt{32}$
   G. $(x + 1)^2 + (y + 1)^2 = \sqrt{32}$
   H. $(x - 1)^2 + (y + 1)^2 = \sqrt{32}$
   J. $(x - 1)^2 + (y + 1)^2 = 32$
   K. $(x + 1)^2 + (y - 1)^2 = 32$

119. In the figure below, line ℓ is parallel to line k. Transversals m and n intersect at point P on ℓ and intersect k at points R and Q, respectively. Point Y is on k, the measure of $\angle PRY$ is 140°, and the measure of $\angle QPR$ is 100°. How many of the angles formed by rays ℓ, k, m, and n have measure 40° ?

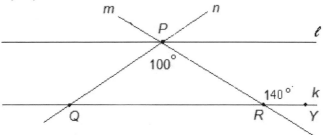

A. 4
B. 6
C. 8
D. 10
E. 12

120. A wheel 37 centimeters in diameter rolls along a straight road without slipping. How many centimeters does the wheel roll along its path in 32 revolutions?

F. 592
G. 1184
H. 592π
J. 2368
K. 1184π

LEVEL 4: PROBABILITY AND STATISTICS

121. An integer from 37 through 842, inclusive, is to be chosen at random. What is the probability that the number chosen will have 9 as at least one digit?

A. $\dfrac{76}{403}$

B. $\dfrac{153}{806}$

C. $\dfrac{77}{403}$

D. $\dfrac{5}{26}$

E. $\dfrac{6}{31}$

122. An urn contains a number of marbles of which 63 are green, 15 are purple, and the remainder are orange. If the probability of picking an orange marble from this urn at random is $\frac{1}{3}$, how many orange marbles are in the urn?

    F. 13
    G. 26
    H. 39
    J. 78
    K. 244

123. The average (arithmetic mean) of x, $2x$, y, and $4y$ is $2x$, what is y in terms of x ?

    A. $\frac{x}{4}$
    B. $\frac{x}{2}$
    C. x
    D. $\frac{3x}{2}$
    E. $2x$

124. A pet store has a white dog, a black dog, and a grey dog. The store also has three cats – one white, one black, and one grey – and three birds – one white, one black, and one grey. Jonathon wants to buy one dog, one cat, and one bird. How many different possibilities does he have?

    F. 3
    G. 6
    H. 9
    J. 12
    K. 27

LEVEL 4: TRIGONOMETRY

125. Whenever $\frac{\tan x}{\cos x}$ is defined, it is equivalent to:

    A. $\cos x$
    B. $\sin x$
    C. $\frac{1}{\cos x}$
    D. $\frac{1}{\sin x}$
    E. $\frac{\sin x}{\cos^2 x}$

126. A 7 foot ladder is leaning against a wall such that the angle relative to the level ground is 70°. Which of the following expressions involving cosine gives the distance, in feet, from the base of the ladder to the wall?

F. $\dfrac{7}{\cos 70°}$

G. $\dfrac{\cos 70°}{7}$

H. $\dfrac{1}{7 \cos 70°}$

J. $7 \cos 70°$

K. $\cos(7 \cdot 70°)$

127. In $\triangle ABC$ shown below, the measure of $\angle B$ is 70°, the measure of $\angle C$ is 60°, and $\overline{BC}$ is 10 inches long. Which of the following is an expression for the length, in inches of $\overline{AB}$?

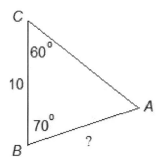

A. $\dfrac{10 \sin 50°}{\sin 70°}$

B. $\dfrac{10 \sin 60°}{\sin 50°}$

C. $\dfrac{10 \sin 130°}{\sin 70°}$

D. $\dfrac{10 \sin 130°}{\sin 50°}$

E. $\dfrac{10 \sin 60°}{\sin 130°}$

128. The vertex of $\angle P$ is the origin of the standard (x, y) coordinate plane. One ray of $\angle P$ is the positive x-axis. The other ray, $\overrightarrow{PQ}$, is positioned so that $\cos A < 0$ and $\tan A > 0$. In which quadrant, if it can be determined, is point Q ?

F. Quadrant I
G. Quadrant II
H. Quadrant III
J. Quadrant IV
K. Cannot be determined from the given information

LEVEL 5: NUMBER THEORY

129. What is the sum of the first 50 terms of the arithmetic sequence in which the 3rd term is 10 and the 15th term is 22?

A. 125
B. 275
C. 525
D. 875
E. 1625

130. In the equation $\log_3 63 - \log_3 7 = \log_6 x$, what is the real value of x?

F. 7
G. 9
H. 36
J. 63
K. 199

131. For all real numbers x, y, and z, with $z \neq 0$, such that the product of x and y is z, which of the following expressions represents the sum of y and z in terms of x and z ?

A. $x + z$

B. $xz + z$

C. $z(x + z)$

D. $\frac{xz+z}{x}$

E. $\frac{x}{z} + z$

183

132. The first and second terms of a geometric sequence are k and r, in that order, where k and r are positive real numbers. What is the 200th term of the sequence?

   F. $(rk)^{200}$

   G. $(rk)^{199}$

   H. $k\left(\frac{r}{k}\right)^{199}$

   J. kr^{200}

   K. kr^{199}

133. Which of the following is true for all consecutive integers j and k such that $j < k$?

   A. j is even

   B. k is even

   C. $k - j$ is even

   D. $k\text{^}2 - j\text{^}2$ is even

   E. $j^2 + k^2$ is odd

134. If $a_k = 5 + 5^2 + 5^3 + 5^4 + \cdots + 5^k$, for which of the following values of k will a_k be divisible by 10?

   F. 37
   G. 51
   H. 75
   J. 88
   K. 91

135. The sum of an infinite geometric series with first term a and common ratio r with $-1 < r < 1$ is given by $\frac{a}{1-r}$. If the sum of a given infinite geometric series is 100 and the first term is 10. what is the value of r ?

   A. 0.1
   B. 0.3
   C. 0.5
   D. 0.7
   E. 0.9

184

136. For positive real numbers x, y, and z such that $3x = \dfrac{y\sqrt{2}}{3} = \dfrac{z\sqrt{2}}{3.1}$, which of the following is true?

    F. $x < y < z$
    G. $x < z < y$
    H. $y < x < z$
    I. $y < z < x$
    J. $z < y < x$

LEVEL 5: ALGEBRA AND FUNCTIONS

137. In the equation $x^2 - bx + c = 0$, b and c are integers. The solutions of this equation are 2 and 3. What is $b - c$?

    A. -11
    B. $- 1$
    C. 1
    D. 5
    E. 11

138. Consider the functions $f(x) = \sqrt{x - 1}$ and $g(x) = ax + b$. In the standard (x, y) coordinate plane, $y = f(g(x))$ passes through $(0, -1)$ and $(2,3)$. What is the value of $a + b$?

    F. 1
    G. 2
    H. 4
    J. 5
    K. 6

139. When $x \neq 7$, $\dfrac{3x}{x^2 - 49} + \dfrac{3x}{7 - x}$ is equivalent to:

    A. $\dfrac{-3x^2 - 21x}{x^2 - 49}$

    B. $\dfrac{-21x}{x^2 - 49}$

    C. $\dfrac{6 - 21x}{x^2 - 49}$

    D. $\dfrac{-3x^2 - 18x}{x^2 - 49}$

    E. $\dfrac{-3x^2}{x^2 - 49}$

140. What is the solution set of $|3x - 2| > 4$?

    F. $\{x | x < 2\}$
    G. $\{x | x > 2\}$
    H. $\{x | x < -\frac{2}{3} \text{ or } x > 2\}$
    J. $\{x | x < -2 \text{ or } x > 2\}$
    K. the empty set

141. In the standard (x, y) coordinate plane, what are the coordinates of the center of the circle whose equation is

$$x^2 - 8x + y^2 + 10y + 15 = 0 ?$$

    A. $(4,5)$
    B. $(4, -5)$
    C. $(-4,5)$
    D. $(-5, -4)$
    E. $(5, -4)$

142. If $f(x) = g(x) - h(x)$, where $g(x) = 8x^2 + 13x - 17$ and $h(x) = 8x^2 - 9x + 16$, then $f(x)$ is <u>always</u> divisible by which of the following?

    F.   5
    G.   7
    H.   9
    J.   11
    K. 13

143. For any real number c, the equation $|x - c| = 3$ can be thought of as meaning "the distance from x to c is 3 units." How far apart are the two solutions for $|x + 2| = 3$?

    A. c
    B. $2c$
    C. $c + 3$
    D. $\sqrt{3^2 + c^2}$
    E. 6

144. For all x in the domain of the function $\dfrac{x+2}{x^3-x}$, this function is equivalent to:

    F. $\dfrac{1}{x^2-1} + \dfrac{2}{x^3-x}$

    G. $\dfrac{2}{x^2-1}$

    H. $\dfrac{1}{x^2-1}$

    J. $\dfrac{1}{x-1}$

    K. $\dfrac{1}{x+1}$

LEVEL 5: GEOMETRY

145. In the figure below, all 4 circles are congruent and each circle is tangent to each circle adjacent to it, and two sides of the square. The area of each circle is 2π inches. What is the length, in inches, of each side of the square?

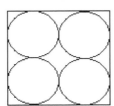

    A. $\sqrt{2}$
    B. 2
    C. $2\sqrt{2}$
    D. 4
    E. $4\sqrt{2}$

146. A square, X_1, has a perimeter of 20 inches. The vertices of a second square, X_2, are the midpoints of the sides of X_1. The vertices of a third square, X, are the midpoints of the sides of X_2. This process continues indefinitely, with the vertices of X being the midpoints of the sides of X_k for each integer $k > 0$. What is the sum of the areas, in square in., of $X_1, X_2, \dots$?

F. $\dfrac{20}{3}$

G. 20

H. 35

J. 50

K. 100

147. In the circle shown below, chords $\overline{AC}$ and $\overline{BD}$ intersect at E, which is the center of the circle. The measure of minor arc AD is 70°. What is the degree measure of $\angle BAC$?

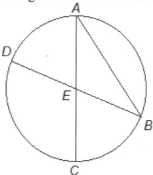

A. 15°
B. 25°
C. 35°
D. 45°
E. 55°

148. Suppose that quadrilateral $PQRS$ has four congruent sides and satisfies $PQ = PR$. What is the value of $\dfrac{QS}{PR}$ to the nearest tenth?

F. 1.5
G. 1.6
H. 1.7
J. 1.8
K. 1.9

188

149. In the triangle below, $DC = 3$ and $BC = 6$. What is the value of AC ?

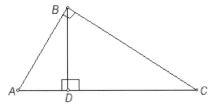

- **A.** 4
- **B.** 6
- **C.** 8
- **D.** 10
- **E.** 12

150. In the figure below, each of the four large circles is tangent to two of the other large circles, the small circle, and two sides of the square. If the diameter of each of the large circles is 10, what is the <u>radius</u> of the small circle?

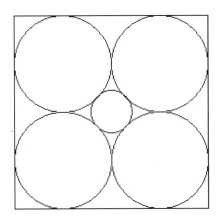

- **F.** $5\sqrt{2}$
- **G.** 5
- **H.** $5\sqrt{2} - 5$
- **J.** $\frac{5}{2}$
- **K.** $5\sqrt{2} - 1$

189

151. A circle with center $(-2, -6)$ passes through the point $(0, -3)$. What is the equation of the line that is tangent to the circle at this point?

A. $y = -\frac{2}{3}x + 3$

B. $y = -\frac{2}{3}x - 3$

C. $y = -\frac{3}{2}x - 3$

D. $y = \frac{3}{2}x + 3$

E. $y = \frac{3}{2}x - 3$

152. In the figure below, the circle has center O and radius 8. What is the length of major arc PRQ ?

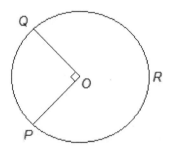

F. 12π

G. $24\sqrt{2}$

H. 6π

J. $12\sqrt{2}$

K. $3\pi\sqrt{2}$

LEVEL 5: PROBABILITY AND STATISTICS

153. A set of marbles contains only black marbles, white marbles, and yellow marbles. If the probability of randomly choosing a black marble is $\frac{1}{14}$ and the probability of randomly choosing a white marble is $\frac{3}{4}$, what is the probability of randomly choosing a yellow marble?

   A. $\frac{5}{28}$

   B. $\frac{3}{14}$

   C. $\frac{1}{4}$

   D. $\frac{2}{7}$

   E. $\frac{9}{28}$

154. The average (arithmetic mean) salary of employees at a bank with A employees in thousands of dollars is 53, and the average salary of employees at a bank with B employees in thousands of dollars is 95. When the salaries of both banks are combined, the average salary in thousands of dollars is 83. What is the value of $\frac{A}{B}$?

   F. .1
   G. .2
   H. .3
   J. .4
   K. .5

155. Two hundred cards numbered 100 through 299 are placed into a bag. After shaking the bag, 1 card is randomly selected from the bag. Without replacing the first card, a second card is drawn. If the first card drawn is 265, what is the probability that both cards drawn have the same tens digit?

    A. $\dfrac{19}{200}$

    B. $\dfrac{19}{199}$

    C. $\dfrac{20}{199}$

    D. $\dfrac{10}{99}$

    E. $\dfrac{24}{199}$

156. A six-digit number is to be formed using each of the digits 1, 2, 3, 4, 5 and 6 exactly once. How many such numbers are there in which the digits 2 and 3 are next to each other?

    F. 30
    G. 60
    H. 120
    J. 240
    K. 720

LEVEL 5: TRIGONOMETRY

157. In the right triangle below, $b > a > 0$. One of the angle measures in the triangle is $\cos^{-1}\frac{b}{\sqrt{a^2+b^2}}$. What is $\csc[\cos^{(-1)}\left(\frac{b}{\sqrt{a^2+b^2}}\right)]$?

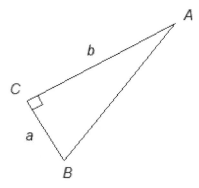

A. $\frac{b}{a}$

B. $\frac{a}{b}$

C. $\frac{b}{\sqrt{a^2+b^2}}$

D. $\frac{a}{\sqrt{a^2+b^2}}$

E. $\frac{\sqrt{a^2+b^2}}{a}$

158. For $0 < x < \frac{\pi}{2}$, the expression $\frac{\cos x}{\sqrt{1-\cos^2 x}} - \frac{\sqrt{1-\sin^2 x}}{\sin x}$ is equivalent to:

F. 0

G. $\cot x$

H. $\tan x - 1$

J. $1 - \tan x$

K. $1 - \cot x$

159. A ladder rests against the side of a wall and reaches a point that is h meters above the ground. The angle formed by the ladder and the ground is $\theta°$. A point on the ladder is k meters from the wall. What is the vertical distance, in meters, from this point on the ladder to the ground?

    A. $(h - k)\tan\theta°$
    B. $(h - k)\cos\theta°$
    C. $h - k\sin\theta°$
    D. $h - k\cos\theta°$
    E. $h - k\tan\theta°$

160. In triangle PQR, the measure of $\angle P$ is $x°$, $PQ = a$ ft, and $PR = b$ ft. Which of the following is the length, in feet, of $\overline{QR}$?

    F. $a\sin 32°$
    G. $b\sin 32°$
    H. $\sqrt{b^2 - a^2}$
    J. $\sqrt{b^2 + a^2}$
    K. $\sqrt{b^2 + a^2 - 2ba\cos x°}$

ANSWERS TO
SUPPLEMENTAL PROBLEMS

LEVEL 1: NUMBER THEORY

1. C
2. K
3. D
4. H
5. A
6. G
7. C
8. J

LEVEL 1: ALGEBRA AND FUNCTIONS

9. E
10. J
11. A
12. J
13. A
14. K
15. C
16. F

LEVEL 1: GEOMETRY

17. E
18. J
19. A
20. H
21. D
22. F
23. A
24. J

LEVEL 1: PROBABILITY AND STATISTICS

25. B
26. H
27. D
28. K
29. D
30. K
31. B
32. J

LEVEL 2: NUMBER THEORY

33. B
34. J
35. B
36. J
37. B
38. J
39. D
40. K

LEVEL 2: ALGEBRA AND FUNCTIONS

41. D
42. K
43. A
44. H
45. D
46. J
47. E
48. G

LEVEL 2: GEOMETRY

49. A
50. H
51. E
52. G
53. A
54. H
55. E
56. J

LEVEL 2: PROBABILITY AND STATISTICS

57. B
58. J
59. B
60. K

LEVEL 2: TRIGONOMETRY

61. E
62. H
63. D
64. G

LEVEL 3: NUMBER THEORY

65. D
66. J
67. E
68. H
69. C
70. J
71. E
72. J

LEVEL 3: ALGEBRA AND FUNCTIONS

73. D
74. F
75. B
76. F
77. E
78. H
79. A
80. J

LEVEL 3: GEOMETRY

81. D
82. G
83. D
84. K
85. C
86. G
87. E
88. J

LEVEL 3: PROBABILITY AND STATISTICS

89. D
90. H
91. C
92. F

LEVEL 3: TRIGONOMETRY

93. E
94. J
95. B
96. J

LEVEL 4: NUMBER THEORY

97. B
98. K
99. D
100. G
101. D
102. H
103. A
104. J

LEVEL 4: ALGEBRA AND FUNCTIONS

105. B
106. J
107. B
108. F
109. B
110. J
111. C
112. K

LEVEL 4: GEOMETRY

113. B
114. J
115. D
116. J
117. C
118. K
119. C
120. K

LEVEL 4: PROBABILITY AND STATISTICS

121. B
122. H
123. C
124. K

LEVEL 4: TRIGONOMETRY

125. E
126. J
127. B
128. H

LEVEL 5: NUMBER THEORY

129. E
130. H
131. D
132. H
133. E
134. J
135. E
136. F

LEVEL 5: ALGEBRA AND FUNCTIONS

137. B
138. K
139. D
140. H
141. B
142. J
143. E
144. F

LEVEL 5: GEOMETRY

145. E
146. J
147. C
148. H
149. E
150. H
151. B
152. F

LEVEL 5: PROBABILITY AND STATISTICS

153. A
154. J
155. B
156. J

LEVEL 5: TRIGONOMETRY

157. E
158. F
159. E
160. K

Congratulations! By practicing the problems in this book you have given yourself a significant boost to your ACT math score. Go ahead and take a practice ACT. The math score you get should be much higher than the score you received on your last practice test.

What should you do to get your score even higher? Good news! You can use this book over and over again to continue to increase your score – right up to a 36. All you need to do is change the problems you are focusing on.

If you are currently scoring less than a 12 you should go back and focus on those Level 1 problems.

If you are between a 13 and 15 you should focus on Level 2 problems, but do all the Level 1 problems and some Level 3 problems as well.

If you are between a 16 and 20, then focus on Level 2 and 3 problems, and throw in a few Level 4 problems every now and then.

If you are between a 21 and 27, then the Level 4 problems are really important. Go ahead and work on all of them, but do some Level 2 and 3 problems as well.

Finally, if you are scoring 28 or higher, it is time to focus primarily on Level 4 and 5 problems.

These are just general guidelines, and you may want to fine tune this a bit by analyzing each of the four subject areas separately. For example, if you are scoring between a 21 and 27 but you are not getting any Level 4 Geometry problems correct, then shift your focus to Level 3 Geometry problems for a little while. Come back to the Level 4 problems after you become a bit more proficient in Level 3 Geometry.

Similarly, if you are breezing through the Level 4 Number Theory Problems, then start focusing on Level 5 Number Theory. On your next practice test you can try those last few Number Theory questions toward the end of each math section.

Upon your next reading, try to solve each problem that you attempt in up to four different ways

- Using an ACT specific math strategy.
- The quickest way you can think of.
- The way you would do it in school.
- The easiest way for you.

Remember – the actual answer is not very important. What is important is to learn as many techniques as possible. This is the best way to simultaneously increase your current score, and increase your level of mathematical maturity.

Keep doing problems from this book for ten to twenty minutes each day right up until two days before your ACT. Mark off the ones you get wrong and attempt them over and over again each week until you can get them right on your own.

I really want to thank you for putting your trust in me and my materials, and I want to assure you that you have made excellent use of your time by studying with this book. I wish you the best of luck on the ACT, on getting into your choice college, and in life.

Steve Warner, Ph.D.
steve@ACTPrepGet36.com

ACTIONS TO COMPLETE AFTER YOU HAVE READ THIS BOOK

1. Take another practice ACT

You should see a substantial improvement in your score.

2. Continue to practice ACT math problems for 10 to 20 minutes each day

Keep practicing problems of the appropriate levels until two days before the ACT.

3. Use my Forum page for additional help

If you feel you need extra help that you cannot get from this book, please feel free to post your questions in my new forum at www.satprepget800.com/forum.

4. Review this book

If this book helped you, please post your positive feedback on the site you purchased it from; e.g. Amazon, Barnes and Noble, etc.

5. Claim your FREE bonuses

If you have not done so yet, visit the following webpage and enter your email address to receive an electronic copy of *The 32 Most Effective SAT Math Strategies*, solutions to all the supplemental problems in this book, and an index of topics with a map to the "Real ACT Prep Guide."

About the Author

Dr. Steve Warner, a New York native, earned his Ph.D. at Rutgers University in Pure Mathematics in May, 2001. While a graduate student, Dr. Warner won the TA Teaching Excellence Award.

After Rutgers, Dr. Warner joined the Penn State Mathematics Department as an Assistant Professor. In September, 2002, Dr. Warner returned to New York to accept an Assistant Professor position at Hofstra University. By September 2007, Dr. Warner had received tenure and was promoted to Associate Professor. He has taught undergraduate and graduate courses in Precalculus, Calculus, Linear Algebra, Differential Equations, Mathematical Logic, Set Theory and Abstract Algebra.

Over that time, Dr. Warner participated in a five year NSF grant, "The MSTP Project," to study and improve mathematics and science curriculum in poorly performing junior high schools. He also published several articles in scholarly journals, specifically on Mathematical Logic.

Dr. Warner has more than 15 years of experience in general math tutoring and tutoring for standardized tests such as the SAT, ACT and AP Calculus exams. He has tutored students both individually and in group settings.

In February, 2010 Dr. Warner released his first SAT prep book "The 32 Most Effective SAT Math Strategies," and in 2012 founded Get 800 Test Prep. Since then Dr. Warner has written books for the SAT, ACT, SAT Math Subject Tests and AP Calculus exams.

Dr. Steve Warner can be reached at

steve@SATPrepGet800.com

BOOKS BY DR. STEVE WARNER

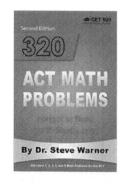

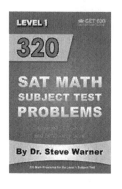

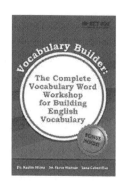

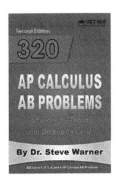

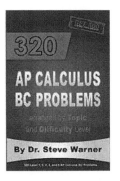

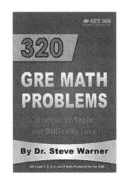

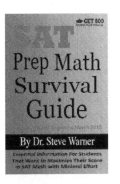

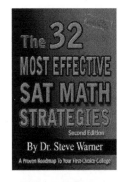

Made in the USA
Middletown, DE
12 April 2017